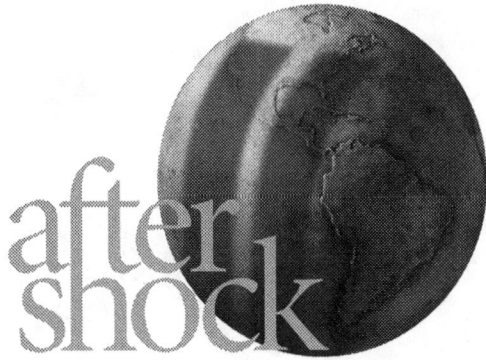
after
shock

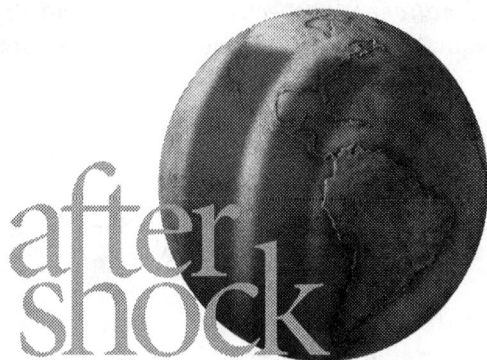

aftershock

RETHINKING THE FUTURE SINCE SEPTEMBER 11, 2001

Pace University, New York City
September 6-8, 2006

www.pace.edu/aftershock

Edited by Katie Hall

PACE UNIVERSITY PRESS   NEW YORK

Also published by Pace University Press
Terrorism and the Psychoanalytic Space:
International Perspectives from Ground Zero
Edited by Joseph A. Cancelmo, Isaac Tylim, Joan Hoffenberg, and
Hattie Myers

© PACE UNIVERSITY PRESS 2007
1 PACE PLAZA
NEW YORK, NY 10038

ISBN 0-944473-83-0

Aftershock : rethinking the future since September 11, 2001 : a
conference held at Pace University, NYC, September 6-8 2006 / edited
by Katie Hall.
    p. cm.
  ISBN 0-944473-81-4 (alk. paper)
  1. September 11 Terrorist Attacks, 2001--Influence--Congresses.
2. September 11 Terrorist Attacks, 2001--Social aspects--New York
(State)--New York--Congresses. 3. September 11 Terrorist Attacks,
2001--Economic aspects--New York (State)--New York--Congresses. 4.
September 11 Terrorist Attacks, 2001--Environmental aspects--New
York (State)--New York--Congresses. 5. Terrorism--New York (State)-
-New York--Prevention--Congresses. 6. Terrorism--Government poli-
cy--New York (State)--New York--Congresses. 7. New York (N.Y.)--
Social conditions--21st century--Congresses. 8. New York (N.Y.)--
Economic conditions--21st century--Congresses. 9. New York (N.Y.)--
Environmental conditions--Congresses. I. Hall, Katie.
  HV6432.7.A3814 2007
  363.325'140973--dc22
                                           2007005290

CONTENTS

Pace University intends this publication to serve as the formal record of the proceedings of "AFTERSHOCK ~ Rethinking the Future Since September 11, 2001," a conference held at Pace University in Lower Manhattan, September 6-8, 2006. All efforts have been made to preserve the spoken character of the panelists' discussions. However, when necessary, the text has been edited for clarity and readability.

FOREWORD

In September 2006, Pace University was honored to host "Aftershock: Rethinking the Future Since September 11, 2001," a three-day conference examining the economic, cultural, environmental, educational, and political consequences of the day five years earlier that forever changed the world.

Participants from across the globe and from our neighborhood near Ground Zero took the occasion to consider the consequences of the terror attacks, lessons learned, and what the future might hold for us.

Our goals were simple: to shed new light on important questions in our post-9/11 world, and to engage in conversation that might lead to a better understanding of that day.

We assembled an extraordinary group of individuals to lead our conversation. More than fifty people on eight panels and four keynote presenters agreed to share their experiences, expertise, and wisdom about the terror attacks. I want to thank each of them, especially keynoters David Gergen, Lee Hamilton, Doris Kearns Goodwin, and William Kristol. Their thought-provoking commentaries gave us a unique perspective on the state of the world five years after the attacks and a context for our panel discussions. Our students and faculty are particularly grateful to them for classroom exchanges during their visits to campus.

Our gratitude is extended to the Pace University Center for Downtown New York (CDNY), which conducted the conference, and the conference director, Meghan French. Thanks to the generosity of donors to CDNY, the conference was presented free and open to the public.

We also thank the Port Authority of New York and New Jersey and Lehman Brothers, whose grants helped underwrite portions of the conference; Crain's New York Business and the Partnership for New York City whose Business Breakfast Forum opened the conference; the New York State Museum for contributing their exhibit, "The First 24 Hours," and television station NY1, which broadcast live from the conference each evening.

We hope our conference made a contribution to the international conversation surrounding the terror attacks, and that this volume will be a resource to scholars and others who will continue to try to make sense of the events of September 11, 2001.

David A. Caputo
President
Pace University

LOWER MANHATTAN

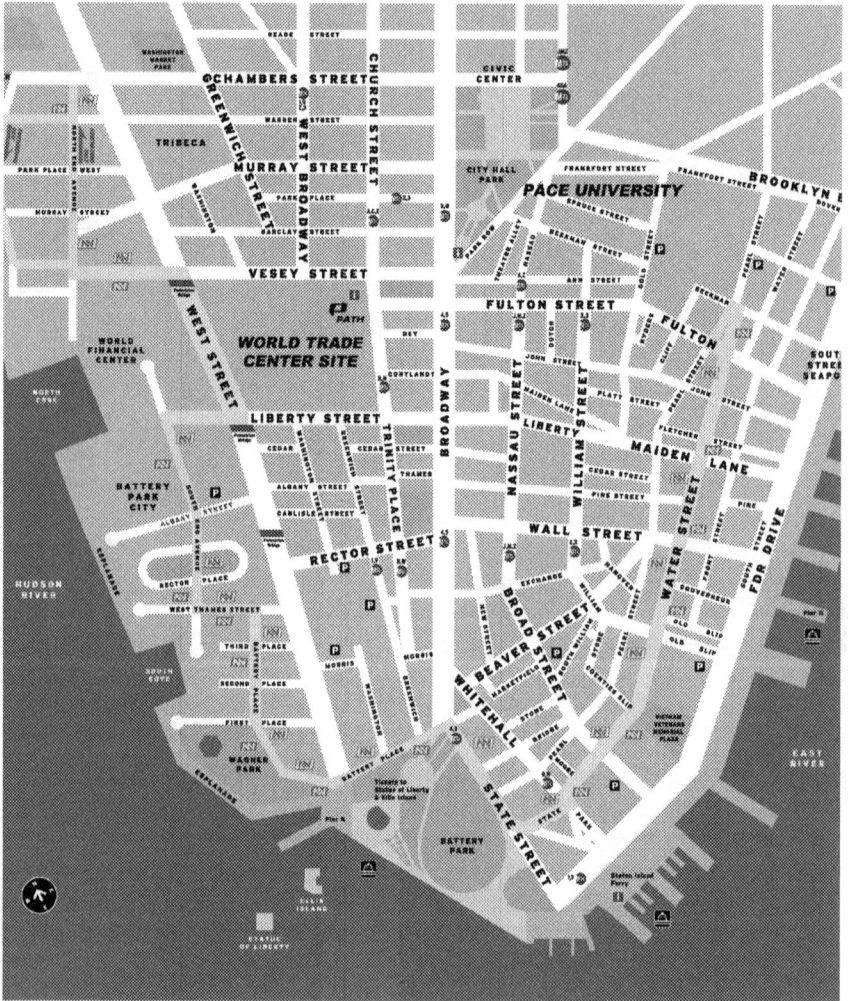

ACRONYMS

ABNY: Association for a Better New York

CBRN: Chemical, Biological, Radiological, Nuclear

BID: Business Improvement District

BMCC: Borough of Manhattan Community College

CB1: Community Board One

CDC: Centers for Disease Control

CICU: Commission on Independent Colleges and Universities

CUNY: City University of New York

DOT: Department of Transportation

EDC: Economic Development Corporation

EMS: Emergency Medical Services

EPA: Environmental Protection Agency

FEMA: Federal Emergency Management Agency

HEA: Higher Education Act

HUD: Housing and Urban Development

HVAC: Heating, Ventilating, Air Conditioning

LMCC: Lower Manhattan Cultural Council

LMCCC: Lower Manhattan Construction Command Center

LMDC: Lower Manhattan Development Corporation

LRA: Louisiana Recovery Authority

MTA: Metro Transit Authority

NAICU: National Association of Independent Colleges and Universities

NY1: New York One

NYCOSH: New York Commission for Occupational Safety and Health

NYSE: New York Stock Exchange

NYU: New York University

OSHA: Occupational Safety and Health Administration
PANYNJ: Port Authority of New York and New Jersey
PATH: Port Authority Trans-Hudson
PCBs: Polychlorinated Biphenyls
PFNYC: Partnership for New York City
RPA: Regional Plan Association
SEVIS: Student and Exchange Visitor Information System
WFC: World Financial Center
WTC: World Trade Center

AN ADDRESS BY NEW YORK CITY DEPUTY MAYOR DANIEL DOCTOROFF AT CRAIN'S BREAKFAST FORUM

There are a *lot* of things you'd think I'd have learned in the last five years. . . .

You'd think I'd have learned not to use the words *MTA* and *rail yards* in the same sentence, but I haven't.

You'd think I'd understand how Albany works . . . but *obviously* I don't.

You'd think I'd have learned not to believe everything I read in the Crain's *Insider*.

Oh, wait . . . I have learned that.

In all seriousness, I'd like to thank Crain's and Pace for organizing this conference to mark what I believe isn't just an anniversary, but also a mid-point of something that began on 9/11 . . . even as we watched and felt the horror of so many terrifying and terrible ends.

We are almost halfway to the completion of one of the greatest comeback stories in the history of New York—the complete re-imagining of Lower Manhattan.

Five years ago we were devastated. We reacted in the ways you expect after such a tragedy: We re-evaluated our loves and lives. We also re-evaluated our city and its place in a world that all of the sudden seemed so ominous. But we didn't retreat from the terror of that day. And why would we? We are a special people, defined by our optimism, itself a product of our collective immigrant experience.

Nearly four out of five of us came from somewhere else—or have a parent who did. It took guts to leave and come to this magical but difficult place. But we did it because we believed in our future here.

So, we didn't retreat. We did just the opposite. We channeled our optimism into a commitment to make our city greater than it was before.

Why else do you think that all over the city, we are only now accomplishing things–big things–that have been talked about for decades?

Like finally realizing the potential of the Hudson Yards, by building our first subway expansion in thirty years, breaking the twenty-year deadlock on Javits expansion, and turning the High Line into one of the city's great parks.

Or bringing major league sports back to Brooklyn on the very same spot that Walter O'Malley wanted to move the Dodgers to 50 years ago. Or by finally turning our decaying waterfront into tens of thousands of homes for New Yorkers of all incomes surrounded by acres and acres of parks. Or stimulating billions of dollars of private investment in the South Bronx —and the thousands of jobs that come with it.

But nowhere has this been truer than in Lower Manhattan. And nowhere was it more important.

Just three days after 9/11, the *New York Times* speculated that "this is the end of the financial district, where the city's history began." Within three months, estimates were that more than 100,000 jobs had been lost . . . with more to come, including Goldman Sachs, which announced plans to shift more workers to a new tower in New Jersey.

But the truth is that fear over the future of Lower Manhattan wasn't new. Lower Manhattan had been in decline almost from the day that Grand Central Terminal opened in 1913. And almost ever since, thoughtful people have been prescribing the same remedy—a highly accessible, twenty-four-hour community where people live and work together, creating a vibrancy that would attract even more corporate tenants.

Here's a quote from the *New York Times* from *before* 9/11:

> Let us face the problem squarely . . . We have neglected valuable opportunities. . . there should be within this area—with all its remarkable possibilities . . . a home building effort . . . then would come places of amusement . . . there is big opportunity for the progressive retail merchant . . . The school question could be easily solved by the city . . .

We should do this, the writer concluded, not only because we *could*, but because Lower Manhattan is different from the rest of New York, different from anywhere.

"There is a spirit," he wrote, "not found elsewhere in the world."

The article was written in 1926.

No, Lower Manhattan never lacked possibilities, or even a plan. What it lacked for decades was the will and the leadership. Until 9/11.

The past five years certainly haven't been perfect or pretty or seamless. But that is inherent in resolving deeply felt wounds, interests, and desires. Mayor Bloomberg and Governor Pataki have led a team of thousands, including Stefan Pryor and the Lower Manhattan Development Corporation, the Port Authority, an array of legislative leaders, dozens of city and state agencies, the Downtown Alliance, Community Board One, and some of the most notable architects, designers, and planners in the world, in creating a plan that will fulfill the dreams that so many have had for Lower Manhattan for so long.

And thanks to Senators Schumer and Clinton, we have the resources to make it happen. And it *is* happening. Almost everything, right now.

There's more than $30 billion being invested in just one square mile—the single biggest concentration of construction activity in New York's history. By 2011, almost every street south of Chambers Street will be rebuilt, including the transformation of West Street from a barren highway to an elegant promenade filled with shade and activity. And $10 billion in transit projects will be completed or will be in construction, including improved subway connections . . . new ferry service . . . and a rail link that will connect Lower Manhattan to Long Island and finally end New York's unacceptable distinction of being one of the only great, global cities without direct airport access.

I'm thrilled that both houses of Congress have included $1.75 billion in funding in their tax bills—bills that we expect to be reconciled this fall.

But it's not enough to make sure people *can* come. We are creating a place where they *want* to come.

In 2002, we predicted that 10,000 new housing units would be created in 10 years. Since then, more than 10,000 units have either opened or are under construction. By 2011, there will be roughly 17,000 new units; and nearly 70,000 people will live downtown—double the number that lives here today, including a major new development at Greenwich Street South. Two new schools and a library will serve the needs of an increasingly family-oriented community. And new shops, restaurants, and cafes are following them, making Lower Manhattan one of the hottest retail markets in the city.

Today it's BMW, Tiffany's (just steps away from its original location on Wall Street), Hermes, Barnes & Noble, Whole Foods, and thirty-one new restaurants. By 2011, more than 800,000 square feet of new retail will open in Lower Manhattan. And thanks to a $38 million investment, Fulton Street, Lower Manhattan's only river-to-river street, will be transformed into a worthy connection between one of the world's great retail destinations at the World Trade Center site and a reborn South Street Seaport.

More than $1 billion is being invested by the public and private sectors in more than 60 cultural institutions or projects—everything from dance, to art, to theatre, to alternative circus performances, plus, of course, the World Trade Center Memorial and Museum, which is destined to be one of the most visited sites in the world.

By 2011, new or rebuilt parks will offer residents and workers an endless array of recreational opportunities or places to relax. A ring of green will wind around the tip of Lower Manhattan and a network of more than 20 new and improved parks are already being built along the narrow streets. Brooklyn Bridge Park and Governors Island will just be a short ferry (or gondola) ride away.

When businesses make a location decision they look for a place that is attractive and filled with activity. The commitments to housing, transportation, retail, cultural institutions, and parks (plus some valuable incentives) are giving them the confidence to invest in Lower Manhattan.

Already, the results are encouraging. As was reported last week, the vacancy rate downtown is poised to drop below 10% for the first time since 9/11.

Some politicians, who sat on their hands for years before 9/11 while Lower Manhattan slowly declined, will say "it's just market forces that are driving the gains."

But look at it a different way:

In 2002, Northern New Jersey's vacancy rate was 2% lower than Lower Manhattan's. Today, it is nearly twice as high. As demand grows, we will have the space—spectacular space—ready. While it wasn't easy getting to this point, by 2011, 12.5 million square feet of new Class A space will be completed or will be ready for occupancy within a year.

Larry Silverstein is so confident in the future that he's already raising the rent for the space the city will take in Tower 4!

Nearly every inch of Lower Manhattan is being rebuilt, reinvented, or reused and most of it will be done by the time we commemorate the 10th anniversary of 9/11.

We all know that so much change in such a small area over such a short period of time will be painful for many. The streets are already congested, small retailers are suffering. That's why the Mayor and the Governor created the Lower Manhattan Construction Command Center, ably led by Charlie Maikish. Its goal is to make sure that we achieve our grand ambitions with as little disruption as possible.

Five years from now, we'll have a chance to measure our progress again. I think we should be judged by the standard demanded by that writer in 1926: to create something imbued with a "spirit not found elsewhere in the world."

And we will succeed. Lower Manhattan will be reborn as something the world has never seen before, and certainly didn't exist before 9/11.

In 2011, Lower Manhattan will be *stunning*. Think about it. Where else will you be able to see the work of seven of the world's greatest architects—Calatrava, Gehry, Childs, Liebskind, Rogers, Nouvel, and Foster—standing side by side?

Lower Manhattan will be *global*. Goldman Sachs, Merrill Lynch, American Express, the New York Stock Exchange, and AIG, plus dozens of other international companies will ensure that Lower Manhattan remains a global center of commerce.

Lower Manhattan will be *for everyone*. Strollers on Wall Street. Picnickers at the World Financial Center. Businessmen kayaking during lunch off South Street. And grandparents cycling on the West Street promenade. Lower Manhattan will never cease to surprise.

Lower Manhattan will be *entertaining*. A visitor will be able to spend the morning at the new National Sports Hall of Fame, the afternoon attending a matinee at the Performing Arts Center at the World Trade Center site, the evening at a classical concert on Governors Island, and then have a drink at a pavilion set against the backdrop of the Brooklyn Bridge and the glittering East River.

Lower Manhattan will be *historic*. As you go shopping inside the renovated Battery Maritime Building or step across the cobbled streets of the Seaport to eat at new restaurant along the river, you'll see how Lower Manhattan will re-imagine its history, retaining the architecture, majesty and intimacy that stretches back to the very beginning of our city while creating something completely modern.

Lower Manhattan will be *connected*. Commuters that today scramble through the warren of staircases and corridors underneath Fulton Street will find new, clear passageways, and even experience a glimpse of sunlight. A trip once filled with dispiriting delays will become dazzlingly simple. And those new easier connections—between transit lines, new ferry stops, and even our airport—will emerge all over Lower Manhattan, creating one of the most accessible districts anywhere.

Lower Manhattan will be *inspiring*. And even as life surges across the neighborhood, restoring a vitality and diversity not seen for a century, we will be drawn to the memorial, to remember. Every day, thousands will come to pay respect to the lives and dreams that ended that day and to be re-inspired, to remind themselves that the only way to truly honor their memories is to channel loss into strength.

And most of all, Lower Manhattan will be *aspiring*. Wherever you stand across New York, you will see the Freedom Tower rising. It will become a compass, the way the Twin Towers once oriented New Yorkers, helping us to understand not only where we are in the city but where our city is in the world. It will become the symbol of New York's unconquerable spirit, the way the tip of Manhattan has always been a beacon of hope and possibility to the millions of immigrants who entered America through its harbor, to the dreamers who made it the site of the world's tallest building–nine times.

Lower Manhattan has always been an embodiment of New Yorkers' ability to imagine something better–and act on it. That's why rather than recoiling from a world that's proven to be more ominous after 9/11, we have had the opposite reaction.

We dug deeper.

Over the next few years, Lower Manhattan will be *totally transformed* because New Yorkers believe in its future today.

And as we take that leap—to leave what we know, in search of something greater—every grand dream and long-delayed vision becomes possible.

PANEL ONE

HOW HAS 9/11 CHANGED THE PREPAREDNESS OF FIRST RESPONDERS: ARE WE READY FOR THE NEXT ATTACK?

JOSEPH RYAN: Good afternoon everyone, I am Dr. Joseph Ryan. I am the Chair of the Criminal Justice program here at Pace University. Prior to coming to Pace University, I was with the New York City Police Department for 25 years. After I left the Police Department I worked for the Justice Department where I helped develop security plans for the 1996 Summer Olympics. While here at Pace University, I was awarded a grant to evaluate President Clinton's hiring of the 100,000 community police officers and I'm currently working with the Department of Homeland Security to develop public-private partnerships that deal with terrorism as well as to promote higher education.

ED GALEA: I'm Ed Galea. I'm the Director of the Fire and Safety Engineering Group at the University of Greenwich in London. I run a group of some 30 people and we spend our time trying to understand how fire develops in structures and spreads, as well as how people react to fire. The reason we do this is so that we can develop and improve our building codes around the world and so we can design buildings for real people, rather than the fantasies that engineers have in their minds. We are also working to improve our computer modeling tools that engineers and architects use to design buildings.

My team is made up of a multi-disciplinary group of people. We have behavioral psychologists, engineers, mathematicians, and computer scientists. Only when you bring a multi-disciplinary team like this together can you truly understand the immensely complicated issues that frame how people respond to particular situations in an evacuation process.

One of the major projects the Fire and Safety Engineering Group is working on is the study of the World Trade Center evacuation. The UK government, through the Engineering and Physical Sciences Research Council, which is the equivalent of

9

the NSF here in the United States, has given us a three-million dollar grant to understand the complexities of the World Trade Center evacuation. We're trying to interview 1,000 survivors to better understand what they went through. What formed their decision making? What hindered their evacuation? What aided their evacuation? And believe it or not, there are lots of fundamental questions that we still don't have answers to. Yet we are still building 100 story buildings all over the world. We don't understand fairly basic questions about how people evacuate and how building structures and the procedures for evacuation that we have in place can aid the evacuation process. That's what I do and that's why I'm here in New York.

MICHAEL EMMERMAN: My name is Mike Emmerman. I'm Director of the Special Operations Support group. I don't do anything technical, obviously; I'm a first responder. I hold positions with the NYPD, the Fire Department, and I'm the government liaison officer with American Red Cross, Today, however, I'm not representing any of those agencies; I'm representing myself.

I was there the morning of 9/11, trapped. That's why I'm here.

JIM DWYER: My name is Jim Dwyer. I'm a reporter with the *New York Times* and have been covering the World Trade Center, and the attempts to knock it down, since 1993.

Along with hundreds of my colleagues at the *Times*, I covered the events of September 11 and the things that went on afterwards. I was part of a group that tried to figure out what happened inside the towers that day. We discovered that there was a great deal of misinformation, misunderstanding, and just plain ignorance about what had happened inside those two buildings. Who escaped, how they escaped, why they escaped; who didn't; and how many people lived through the crashes but couldn't get out of the buildings—these seemed to be fundamental questions that needed to be answered, so a group of us worked for a better part of a year on a series of stories.

We sued the City of New York and the Port Authority of New York and New Jersey. Though we're still in court with them, we've managed to get about 20,000 pages of printed records and hundreds of hours of tapes released that provide very concrete information about what went on. We've also interviewed several hundred people who were inside the towers that day. With one of my colleagues, Kevin Flynn, I wrote a book called *102 Minutes: the Untold Struggle to Survive Inside the World Trade Center*. That's why I'm here.

JOSEPH RYAN: Welcome to Pace. The panel discussion this afternoon focuses on the issue of preparedness and response.

What happened on September 11, 2001? I always like to anchor my discussions on this topic to a document that most of us are fully aware of but often forget about. It's called the U.S. Constitution. It says that "we the people of the United States, in order to form a more perfect union, provide for the common defense." Who was the common defense on September 11? Where was he? I know you're going to say, "who is *he*?" Tom Cruise? Top Gun?

Most of the expectations we have in society come from the media, from television, from movies. Somehow, we were all led to believe that on September 11, Top Gun was going to be there. I don't even want to believe that they were supposed to shoot down those airliners, but the reality is that that's what they were supposed to do. What happened? Where was the Air Force? Where was the Navy that day? There is a video of the Police Commissioner on the morning of 9/11 showing him screaming at his detective, "Call for air support." The detective looked at him, completely confused, "What's the number?"

That's the reality we face when trying to figure out what the 'common defense' is in our system of government. The answer, I believe, is that when we're talking about the common defense, about the general welfare of our people, we're talking about first responders.

I would like to thank each of our participants for agreeing to be here today. I have three broad questions and I'm hoping that somewhere in these panel discussions each of them will be answered.

First, what was the role of first responders prior to September 11, 2001? Second, what has changed for these organizations or individuals since then? And third, what still needs to be done to ensure appropriate response to future terrorist attacks?

MICHAEL EMMERMAN: I think you have to understand, first and foremost, that first responders in New York City—the police, the fire department, and EMS—respond to at least eight "disasters" a day. We have very, very capable responders in all those fields. So the question for today is, "what are we preparing for?"

Post 9/11 we are prepared for most things other than very large scale, broad-based issues. I don't know how many of you remember the hurricane that flooded Annapolis. At that time, I was sitting in the command center watching the weather and listening to the broadcast. We were all waiting for that storm to turn. If it had turned one degree north, Manhattan, up to 14th Street, would have been under water because of the storm surge. Other parts of Brooklyn and Staten Island and Long Island would have been flooded as well. And we didn't have an evacuation plan. It would have gotten really interesting if that had happened. Thank God it didn't.

Fast forward to today. We're in the middle of a supposed eleven year cycle on hurricanes and we still don't have a real evacuation plan. But, there are hundreds of very serious, smart, forward-thinking people who work on this issue every single day. What we don't have—and I am not representing any single agency, but working with everybody—is the money and the people power to implement everything we need to do. It's not a mystery. Everybody knows what needs to be done. The question is, how do we get it done? Where do we get the resources to get it done? Where do we get the people?

I'll give you one story. If we needed to evacuate 750,000 people, it would mean setting up hundreds of shelters. It would mean that each of the agencies I mentioned would need ten thousand volunteers at a moment's notice. We don't have ten thousand volunteers. Right now, we're lucky if we have three or four thousand that are truly committed to the process. Even if we trained ten thousand people to participate at a moment's notice . . . if the entire city of New York were under the threat of a hurricane and people were evacuating, I don't know how many of those ten thousand volunteers we'd lose. Remember the buses in New Orleans that didn't get moved because there were no bus drivers to move them? Who's going to drive the buses to move the people here in New York City? Who's going to drive the trucks?

There are thousands of issues that need to be dealt with and they are being dealt with, I just don't know when they'll be resolved. Going forward, I think we all would like to know what the efficacy of the building is. What are the risks? What are we getting into?

What happened on 9/11 raised several issues. I, as a first responder, would not have been running into the building if I thought that the building was going come down on me. I share that attitude with a whole bunch of other people. We had no idea. Yes, many people were called heroes for running in, but it would have been a great thing to know that the buildings could possibly come down, which we didn't. So, what has changed?

One of the things that changed was our understanding of our communication needs. I've heard arguments back and forth about the radio system—that we all need to communicate with each other on the same channel. I don't know if you've ever been inside a command center or out in the field at a real disaster, but if everyone could talk to each other, that would be a disaster. Every agency has its own command and control. What's required is that every command and control center in every agency can communicate with those of the other agencies. That doesn't mean that I, as a single responder, need to be able to talk to a fireman or a police officer on the same radio. I may need to get instructions, and all three agencies may need [to share] instructions at the same time. But that's rare. This whole argument about radios drives me nuts.

But yes, we now know that command and control needs to communicate better and have the resources to do that. The cell towers failed, communications failed, suddenly the three and a

13

half pagers and the two cell phones that I carried were worthless. So now we know we need a new system to communicate. All of these things need money, and they need people, and that's where we're headed.

JOSEPH RYAN: Thank you, Michael. Since you did raise the issue of buildings, I'm going to shift to Ed Galea. As this is your area of expertise, please share with us some of your insight.

ED GALEA: I think if you want to look at the situation, you need to consider what's expected of our first responders, what's expected of our buildings, and what's expected of our building occupants. If you go back to pre-9/11, not a single high rise building in the world was designed to accommodate a full scale evacuation. All the buildings that stand in New York today? You cannot evacuate the entire population of those buildings in a useful amount of time. The buildings simply are not designed for that. Our building strategies for high rise structures around the world were based on the "defend in place" strategy. When you had an incident on a particular floor, the well-engineered building was meant to contain the incident on that floor. All you would be expected to do is to evacuate the incident floor, the floor above, and the floor below. So our first responders had a relatively easy job. I mean, of course it's not an easy job; they have a very, very difficult job. But if you consider that that's the job they had to contend with, to make sure that those people on those three floors were evacuated, then they could go in, assist those people, and then tackle the incident. That's manageable.

But in the post-9/11 world, we've seen that there is a need to design our buildings so that we can expect people who are in those buildings to evacuate in a period of time that is deemed to be acceptable. That's the world we live in today. Do our buildings currently have that capability? No, they don't. Are we designing our buildings to meet that capability? Well, the building codes don't require it. And if the building codes don't require it, then I'm afraid architects and engineers won't necessarily build that sort of capability into the structures. So in the post-

14

9/11 world, are we better prepared for these types of incidents? I don't think we are.

I've traveled a lot of New York on your underground. Your subway system scares the living daylights out of me. If there was an incident like 7-7, that we had in London, would you be able to evacuate people from the underground stations in New York? Quickly? Efficiently? I don't know. . .I don't think so. With all those steel gratings around the entrances and exits? When I've raised this with people in the Fire Department, they say, "Well, we'll come down and we'll cut the barriers and let the people out." I don't think that's going to happen. I think it's going to be very, very difficult.

Are we better prepared in the post-9/11 world? I think we're beginning to ask questions and I think we're trying to find answers. No one is suggesting that we need to design buildings to withstand a 747 or an Avis 8380 smashing into the side of it. But I believe it's not beyond the wit of man to design buildings so that they will stand up long enough to allow people to get out. That's the key thing. What's the role for our first responders? Well, they're not going to be able to tackle these sorts of incidents, these fires. What I think their role is going to be is to help aid people to get out of the buildings quickly and efficiently. But the real emphasis must be on building design. We have to design our structures to enable people to take appropriate evacuation actions.

JIM DWYER: Joe mentioned before, "Where was Top Gun?" And I will tell you where Top Gun was. Top Gun was at the desk next to you. Top Gun was the woman in Human Resources who got everybody up out of their desks and into the stairwell. Top Gun was the man with the flashlight in his drawer, and a whistle, who walked up a flight of stairs with another man and pried open the doors where people were trapped.

On the top floors of the Trade Center, in both buildings, there wasn't a prayer in the world that a first responder was going to get to them. They were a thousand feet up. It was four hours walking with all the gear a typical firefighter wears. Ninety-seven and a half pounds of coat, boots, helmet, and ax.

15

The first responders got about thirty stories into the building. Some of them got a little higher. But for the most part, the people doing the response were the ones who are sitting here in this auditorium.

So on Ed and Mike's discussions about building design being so important, I'd like to echo that because it is clear that the tools for people to be saved in such a situation must be in the hands of the people who might be trapped. That means in the case of the World Trade Center, had the buildings been completely occupied, had there been twenty-five thousand people in each of those towers, instead of about seven or eight-thousand, given the rate of descent in those buildings, probably fourteen or fifteen thousand people would have died, not 2,749.

We're very fortunate that the attacks came on a day when the buildings were not as busy as they usually are, on an election day, the first day of school, you know, all that stuff that was thought to have impacted the occupancy. And at the time of day that the attacks did take place.

The second thing is that our emergency response culture in New York is both incredibly muscular but also muscle-bound. It, at times, lacks the kind of vision to go with the valor that I think would do us a lot of good. Let me give you an example. I have some transcripts from a police helicopter that was up in the air on the morning of 9/11. I'm going to read a little bit from that, and then tell you a little more about it: Helicopter says around 9:58, (this is just the moment when the South Tower, the first of the two buildings falls), "we just had a collapse of the South Tower. Advise any police, this is aviation."

Then the dispatcher says, "What building's coming down?" Then they go back and forth about which building was coming down.

"Number two has just collapsed." And then one of the chiefs on the ground says, "Give me an eyeball report on the other building." At that point, there's radio traffic, lots of traffic, then the voice comes through.

"Advise everybody to evacuate the area. About fifteen floors down from the top, it looks like it's totally glowing red on the inside, it's inevitable."

16

And the dispatcher says, "All right. From the fifteenth floor down it looks like the building is going to collapse. We need to evacuate everyone from the vicinity."

Another pilot jumps in from another helicopter, "I don't think this has too much longer to go, I would evacuate all the people within the area of that second building." Those are both NYPD transmissions from the NYPD helicopters.

The primary responders in the building that morning were firefighters. Yes, there were NYPD and officers from the Emergency Services Unit and some other officers, but primarily you had, by our estimation, about four hundred firefighters inside of the North Tower at that point.

Some of them heard from their own commanders that they were pulling out. Let me tell you, 9:58 the first building goes down; 10:28 the second building goes down. The question is what happens in those 29 minutes? During those 29 minutes, scattered groups of firefighters heard that it was time to pull back, but many others had no idea whatsoever. It's around the time of those helicopter transmissions.

I'm going to read a little bit here:

> "Around that same time, another group had reached the nineteenth floor on their way down. Three court officers, Bachilari, Mascola, and Wender, who had gotten up to the fifty-first floor, were now turning back. They stepped out of the staircase and into the corridor. They could scarcely believe their eyes. The nineteenth floor was just as full then as it had been when they came up, still packed with firefighters from end to end of the hallway and down other corridors. So tight it would be tough to find a place to squeeze in alongside the wall with them. The place was carpeted with firefighters. Most were sitting and had stripped off their turnout coats. Helmets off, some were down to their blue t-shirts, maps of sweat blotting through the fabric and blazing the Fire Department shield. Officer Wender saw that some were lying down and there were axes leaned against the wall. Legs stretched out, arms resting against oxygen tanks. They could not be hearing, Wender thought, what we are hearing. They

17

guessed that there were at least one hundred fire fighters on the floor. "We're getting out of here," Officer Bachilari yelled. "We've been told we've got to get out of the building." "Alright," said one of the firemen, "we'll come down in a few minutes. We'll be right there."

And at that point, I think it's worth noting that through that entire morning there was no joint command between the NYPD, the FDNY, or the Emergency Medical Service. There was no interoperability between the command and control functions of each of those agencies. So the Chiefs in the Fire Department were not hearing from the Chiefs in the Police Department that the helicopters up in the sky operated by the NYPD [were saying] that this building was going to come down.

Now, it is a unique situation. But it's not hard to imagine other situations where the assets of one agency are finding out important things and there is no easy, reliable mechanism for sharing that information. The situation had existed before 9/11, going back, for sure, to the 1993 bombing. It was a primary focus of the after-action reports that were done then. There actually were radios bought during the 1990s that would allow interoperability. These radios, in most cases, never left the precinct offices or got out of the back of the Fire Department Battalion Chiefs' cars. The technology existed for good reliable communication; the management willpower to make it happen did not.

Embarrassment, I think, hardly begins to describe the emotion, or the pain and regret, over losing hundreds of firefighters who could have been pulled out had they known the building was coming down. I think that has prompted these two agencies to make some efforts to improve their interoperability, their willingness, and their instinct and culture about working together.

JOSEPH RYAN: Ed, you commented about the evacuation process for buildings, and how it traditionally was designed for one floor above, one floor below. What are we supposed to do right now with all the existing stock of buildings in New York City? How do we evacuate?

ED GALEA: One of the things that I'm very pleased has been introduced in New York is that at least you're doing drills, and you're training people to respond. I think until fairly recently you didn't have a requirement for full building turnout so that you could actually practice evacuating an entire building. Now people have the experience of going through that process. I think that's very important. One of the lessons of 9/11 is that so many people, even after the first incident, the '93 bombing, didn't know where the staircases were. They had no idea what it was like in the staircase, had no concept of how high off the ground they were, or how difficult it would be to travel down stairs. The drills and the exercises that were in place were a mockery. You'd just be sitting down at your desk and somebody would tell you, "the staircases are over there, and you've got to go to the staircase." I mean that is just absolutely ridiculous. For such large, complex structures you need to rehearse and you need to rehearse and you need to rehearse. I can't emphasize that enough. You have to train people. You don't want people in an emergency situation, in a life and death situation, to do something for the first time. That's a recipe for disaster.

JIM DWYER: I think, Ed, if I could jump in at that. One of the reasons that it's so important is that the people who went to work in the Trade Center, for instance, were people who made their money not from salaries, but from trades and other operations. So unless it was clear to them that they really needed to leave the building, leaving their desks meant they were going to lose a lot of money. So there were strong incentives to stay put unless you were persuaded that you were in big trouble.

ED GALEA: One of the major problems in evacuation practices around the world is this issue of occupant response time. It's the time from when the occupant is alerted to the fact that they have to evacuate to the time that they actually disengage from their normal activities and engage in the evacuation process. It's called the response time. And prior to 9/11, response time was huge. People just . . . the alarm goes off and people just wouldn't go. Because they're too busy making money, or they think that it's a false alarm or another damn drill. In fact, what you find in a lot of evacuation situations is that the response time is actually longer than the travel time.

Immediately post 9/11, response times were quite short. Alarm goes off, everyone runs out of the building. Now, five years later, what's happening? Well, people are lapsing back into the same old mentality and taking longer to respond. We interviewed some survivors recently. They've moved to another building. And some of the folks who didn't want to talk to us formally in an interview did agree to have lunch with us. They were traders, and I was staggered to learn that even though they survived 9/11, they didn't know where the evacuation stairs were. This is the sort of mentality we've got to fight against. And this is why it's so important to do drills.

What do we do with the current building stock? There's not a lot we can do. We can train the occupants. We can emphasize to them the importance of understanding why it's imperative to evacuate. One of the things that staggered me was the amount of time people spent on their cell phones to call their families on 9/11. They were doing so while they were still in the buildings. They were not safe. They were in a place of danger and they stopped the evacuation process to call their family to say, "I'm okay, I'm evacuating." Well, you don't do that. You're not safe, you don't call anyone until you get out of the building and you're in a place of safety.

This is the sort of simple thing that we've got to reinforce with people.

Another thing that staggered me, a simple thing, was the number of shoes. Women's shoes were scattered throughout the staircases. Why? Well they were these ridiculous torture devices that women wear. These high heeled shoes. You can't evacuate down fifty flights of stairs wearing those shoes. So women took them off and evacuated in bare feet. Dumb idea ladies, don't do that. Because there could be debris, broken glass. If you're going to hurt yourself, you're going to delay your evacuation process. Either wear sensible shoes in the first place or have a sensible pair of shoes under your desk so that when you've got to evacuate, you just slip them on and evacuate. And if you are wearing high heeled shoes, don't throw them away, hold on to them during the evacuation process.

So there are many simple things we can do to improve the evacuability of our current building stock. It's not just simply redesign. It's not, "what do we do with all the old buildings?" It's the people that are important. It's in the processes that we have in place that people can make a big difference in the success or failure of an evacuation. It's not simply the design of the structure. Sure, that's got a big part to play, but the people who use the buildings are probably the most important part of the evacuation system.

MICHAEL EMMERMAN: Prior to 9/11 there were very few organizations in New York that actually had evacuation packets for their employees. Many of the larger corporations now provide them to each and every employee. They're sort of Velcro-ed onto their desks. They contain little things, like an oral/nasal mask or maybe a flashlight. My experience has been, as Ed's has, that the longer we get away from 9/11, the more complacent we get. I work on the thirty-eighth floor of a forty-four story building and I know that there is a small portion of the employee staff that is still frightened and at the first sound of an alarm will leave the building. The majority, however, will sit there until they are *ordered* to leave the building. It takes that kind of urgency to get these people out. And they're still not performing as they should. Evacuation is not just leaving the building. If you work with a large organization or in any building that does this right, there are staging areas outside the building that you should report to. The reason for that is that if you report to a staging area, and someone can mark off that you have exited the building, there are then fewer people to look for. Because if they have a long list of people that they're still looking for, that's when the rescuers and responders come in and need to climb all those flights to find you.

If you are told to go to a staging area, go. Check in with somebody and let them know you're out of the building.

ED GALEA: I want to come back to a point you just made. Even in 9/11 some people actually had packs that their companies had provided them. And they didn't use them. They evacuated without their packs. So it's the training issue again. Training, training, training is the most important thing.

Whenever I travel, I carry a smoke hood with me. When I fly, I've got a smoke hood, when I stay in a hotel, I've got a smoke hood. Unfortunately, on this trip I couldn't bring it along because of the increased security, believe it or not. I feel naked without my smoke hood. And not only do I have a smoke hood, but my wife has one too and we train putting the damn thing on. Because it does no good to have a smoke hood if you've never put it on. These things are complicated.

JIM DWYER: Ed, what is a smoke hood like? Where do you get them?

ED GALEA: The smoke hood I have is a device that folds down to a pack about this big. It's a device that you pull over your head with a large visor that you can see through. There's a rubber seal to prevent the toxic gases coming up into the smoke hood and there's a filtration system. You've got to be careful which smoke hood you buy, because a lot of them on the market are rubbish. If they don't filter out carbon monoxide, you're wasting your time. So you want to make sure that the filtration system you've got is good enough to filter out carbon monoxide and preferably hydrogen-cyanide, as well. They're two of the major killers in fire gasses. The device I have will give me twenty minutes of breathable air. It's something that I practice putting on and I get my wife to practice putting it on. It mucks up her hairdo so she doesn't like it too much. But you need to be able to put these things on quickly and experience what it's like walking around with them on.

MICHAEL EMMERMAN: And I will advise you, that if you are buying one, buy two of them and take one, open up the envelope and try putting it on, practice with it. It doubles the cost, but what the hell? If you are buying it for a family or a group of employees, you're buying a number of them. Buy extra ones that people can actually try on. Because the last thing you want to do in an emergency is stand there figuring out how to get this thing on your head. And it's really uncomfortable, so you have to know that, and know that you *can* breathe, because when you first put it on, there's a sense that you can't. So you need to practice with this stuff. It's not enough to have it, you have to know how it works. And some of them deteriorate over time, so you don't want to open the envelope of the primary. You buy another one and practice with that one.

JOSEPH RYAN: Jim, in the narratives you just shared with us, those powerful words from individuals who were trapped should've been monumental lessons learned. Yet I just heard Ed say that complacencies are coming back, response times are beginning to increase. How would you go about helping us get the message out?

JIM DWYER: I think you're going to see plenty of coverage of where we are five years later. That coverage is going to run the gamut from cultural and social issues and clinical stuff to behavioral things like increasing complacency. I think these anniversaries do make us revisit things that we may not want to think about, but they also have the benefit of reminding us of others. You know I'm a pretty well-informed guy and I didn't know about smoke hoods, so I'm glad to hear Ed and Mike talk about them. That might be the kind of thing that you'll end up seeing in the newspaper.

ED GALEA: Just to follow up, I've only just come back from a business trip to Korea and I was pleased to find that in my hotel room, the hotel had provided a smoke hood, a strong flashlight, and a harness cable that you could attach and lower yourself out the window if you had to. I really wouldn't want to do that, but all of this equipment was provided in the hotel room. Unfortunately, there were no instructions on how to use any of the stuff, which is probably more dangerous than not having it

23

there in the first place. But they're moving in the right sort of direction.

JIM DWYER: But in American hotels, you get HBO.

JOSEPH RYAN: Michael, one further question, then we'll open up to the audience. You mentioned a figure of ten thousand first responders being trained. Who's doing the training? How effective is the training?

MICHAEL EMMERMAN: There need to be ten thousand in each particular group. Let me give you a little bit of background.

In classic terms, the Police Department and Fire Department are always under-staffed. Period. That's before 9/11 and post 9/11; it's just what it is. And that's true in most cities of the United States; we're not an exception. Even when you go to the other volunteer groups, like the American Red Cross, they never ever have enough volunteers who are trained. Immediately post 9/11, there were, I think, something like fifteen thousand people standing in front of the Red Cross building at 150 Amsterdam Avenue in New York City, all wanting to volunteer. They wanted to do something. But there was very little that they could do because they weren't trained. Spontaneous responders are almost always useless and in some cases they can make it worse. So today we need ten thousand volunteers just for the Red Cross, trained volunteers to handle an evacuation of 750,000 people. That's just on the Red Cross piece, it's not including the Police Department, or the Fire Department, or EMS. It's just that one piece. Today in New York City we have approximately four thousand properly trained volunteers that we can count on. So, the training is getting done. We're out there trying to find people to add to that cavalry, but this is a very slow process. And you will, over time, lose people unless you use them. So, that's the ten thousand.

JOSEPH RYAN: I'll just throw in one comment. I just returned from a conference sponsored by Homeland Security and Ed was talking about training and training and training. Homeland

Security guestimates now, with FEMA, that there is a need for eleven million first responder education training programs to go on. And when you mention American Red Cross, all these volunteers lined up to help, but no one knows how to help.

Let me ask the audience. Are there any questions for the panelists?

AUDIENCE: I'm curious about why you have not talked about a radiological disaster, and what's necessary to cope with that, for both first responders and the public. Thank you.

MICHAEL EMMERMAN: I love this one. Okay, you have nuclear, biological, chemical, and improvised explosive devices. You have hurricanes, fires, and floods, and a bunch of other disasters. The least likely is radiological and the hardest to prepare for is radiological. You work on a scale of most likely, most probable, when you're trying to train people to prepare. Preparing for a radiological disaster requires an enormous amount of money and equipment. I wish we had it, I actually do. And I wish we could train more than one elite team in a location on exactly what to do and how to prepare for that. The fact remains that the money that's spent within a bureaucracy, within a democracy, is spent on a scale from what's most likely to happen to what's least likely, and what's going to happen soonest.

ED GALEA: I think the CBRN scenarios—that's Chemical, Biological, Radiological, and Nuclear—are truly nightmare scenarios. We've been training our responders to address these situations, but if the bad guys can actually deliver weaponized chemical agents in our big cities, in our underground systems, I think we're going to have a really big problem. We've been training our first responders to cope in London. We've equipped our fire services with CBRN equipment suits, we've instigated decontamination units, last year we had a major exercise in the city of London where we actually exercised our joint forces—our police and our fire fighters and ambulance services—to respond in fully turned-out equipment. And we learned a lot from that.

It didn't go very well, but we learned a lot. We need to do more of that, but there are a number of issues that we still need to address. And not the least of which is that you can have a first responder in the correct equipment, but the physiological assault on the first responder, when they're wearing this CBRN kit, is fairly dramatic. How much can they do before they have to disengage and cool down? Re-hydrate and cool down are big issues. As I say, it's one of the nightmare scenarios.

MICHAEL EMMERMAN: May I add one thing to that? We've had several drills here in the New York City area that you may not be aware of where we do this similar kind of work with all of the agencies. My issue is that we just don't have enough of them and we just don't have the money to support a broad scale kind of training.

AUDIENCE: We've heard there are a lot of problems, that there's not enough money, and not enough people to respond. But what if a disaster does come? Something like 9/11, or something like Hurricane Katrina? What could be done? Is the only answer that just more people will be lost, just because there's not enough money and volunteers?

MICHAEL EMMERMAN: That's a really tough one. The bottom line is that if there were a very broad scale issue, like a nuclear attack, there would be incredible chaos. No one has a concept of how that would look. The fact remains that in a situation like that, we in New York City, though we normally stand on our own, would be getting help from the outside dramatically. We would need it. And it would take days to get. Even if everyone was all set to go, it would take days. There is no easy answer to that. Even in New York City, if we had all of the numbers of people that we needed and all of the equipment and all of the training, having been involved with Emergency Response for now twenty-seven years, I will tell you that something's going to go wrong, period. It's not going to work the way you thought it was going to work. So, I don't have an easy answer for that.

JOSEPH RYAN: I would just like to add a sense of optimism. Homeland Security is less than five years in existence and they are constantly learning lessons. One of the lessons that we have learned, which Ed alluded to, is training, training, training. We have to get the message out. In the next couple days, as Jim was saying, there'll be tons of media coverage on what has happened since 9/11, and what we have learned. I think we all need to listen. Just be optimistic

JIM DWYER: Don't be afraid. We can live through disasters, people do. In the World Trade Center, 99 percent of the people below the point where the planes hit, escaped and survived. They didn't have training, they didn't have equipment, they had good sense and they took care of each other and they made sure they got out. You're going to be able to do that if the time comes for you. Have faith. Don't let fear paralyze you. There's too much fear-mongering in this country.

AUDIENCE: It's my understanding that the city fire codes and building codes did not apply to the World Trade Center. Are there any folks here who can talk about whether they did apply, and how much more time would have been gained before the collapse happened had the buildings been up to code?

ED GALEA: I don't know the New York code well enough. You're talking about the structural elements? The failure in the design to cope with the type of incident would have been the same even if the buildings had followed the New York code, in terms of the structural components. I'm not sure about the fire-proofing, if there would have been more requirements to have better fire-proofing, but I don't think it would have made a lot of difference. There are some fundamental problems with the way the building was designed, which I think would have been committed even on the New York code. For example, the staircases, in my opinion, were too closely [positioned]. If you're going to have one cataclysmic event that's capable of taking out the staircase, it's likely all three of them would have been taken out. And so there were some fourteen hundred people who lost their lives simply because a staircase didn't remain intact from top to bottom, in the North Tower at least.

Another issue, I think, is the construction of the staircases. I'm not sure if New York code required them to be made out of concrete. They were built, essentially, out of plasterboard. At least that is what we call it in the UK. And really, that's a good, fireproof material, but it has no resistance to blasts or that type of incident. I don't know if that would have been any different under the New York building code.

The other point is that, as Jim pointed out, had the buildings been fully occupied with twenty-five thousand people . . . we've simulated the evacuation process under that scenario and it would have taken, even with the three staircases intact, almost three hours to get those people out. From our calculations, something like eight thousand people would have lost their lives in the North Tower alone and that's if the staircases had remained intact.

JIM DWYER: Let me just say a couple things about the code. One of the most shocking things I learned in my research was how paper-thin and meaningless the codes ultimately are. They're not based on real-world situations. Let me give you an example. They're supposed to fire-test the metal that goes into these buildings, the steel. They had to test a seventeen-foot length of steel to see if it could stand up for a certain number of minutes to a blast furnace. Well they found out that a seventeen-foot piece of steel could. Here's the catch: they didn't make the Trade Center with 17 foot pieces of steel. They made it with thirty-five-and fourty- foot pieces of steel. The difference is substantial. The fire tests that were done on the true length of the steel weren't done until 2003 or 2004, after the buildings had fallen down. And the steel stood for the amount of time it was coded to stand.

The Port Authority took the position in 1965 that they were going to meet or exceed the New York building codes, even though they weren't required to. In fact, they did not meet the codes in all respects. Most importantly, it turned out they were short one staircase. They were supposed to have had four stair- cases in the building and they only had three. Had the fourth staircase been in the building, in each of the buildings, had it been out of the line of impact, no doubt more people would have survived.

Okay, so that's the end of the code.

JOSEPH RYAN: Ed, Michael, and Jim, on behalf of President David Caputo and the Pace University community, thank you for coming.

PANEL TWO
ECONOMIC IMPACT: GLOBAL BUSINESS COMMUNITY

JOE BACZKO: Good afternoon ladies and gentlemen. Welcome again to Pace and to this afternoon's panel. My name is Joe Baczko and I'm the Dean of the Lubin School of Business here at Pace University. It's a great pleasure to be a Dean at a school that has been around for 100 years.

I spent the last thirty years in international business. What better place to lead a Business School than in the global center of business?

Here at Pace, we have five thousand students in graduate and undergraduate programs. More importantly, they represent over sixty different nationalities and make this university and this school very much a part of New York City.

We've been an integral part of the tapestry of New York City, and more importantly downtown New York, for a century. Since September 11, 2001, we've been measuring the economic well-being of Lower Manhattan through the Pace Downtown Index. For the last twelve to fourteen, maybe even sixteen months, the index has shown that downtown New York—Lower Manhattan—has been growing at a relatively healthy rate, about 3.7 percent compounded. That's good news, a good sign.

Our topic today is the economic impact of the terror attacks of 9/11 on the global business community. There is probably no better panel to discuss this issue than the one that we have with us. I'll just introduce them very briefly, and then they will all do their own introduction later on.

Eric Deutsch is the President of the Alliance for Downtown New York; Michael Dolfman is the Regional Commissioner of the U.S. Bureau of Labor Statistics; Joe Petro is Executive VP and Managing Director of Citigroup Securities and Investigative Services; and Steve Spinola is President of the New York Real Estate Board.

On September 10, 2001, New York City was already in a serious recession. The stock market, particularly the tech side, had tanked. We had wiped-out savings and endangered 401k plans long before Enron. The economic bust was pervasive. Real estate values were headed down. Jobs were disappearing, particularly high paying jobs, and even our standing in the world in terms of goodwill was on a downturn.

Recently, I went back and looked at the *New York Times*, the *Wall Street Journal,* and the *Washington Post* three weeks after 9/11. There was still an awful lot of coverage about everything that was happening here in New York, in terms of the disaster, but overwhelmingly the predictions going forward were pretty gloomy. Essentially, people believed nothing would ever be the same as it was before 9/11. Globalization would be curtailed. The New York downtown, if not all of New York, would suffer a permanent exodus which would change it as a global center of commerce. Tourism, international and domestic, would disappear.

It was feared that students—and for us that was a very important thing—whether they were domestic or international, would go elsewhere. Those who were heading up this administration at Pace, including President Caputo, had to be very, very seriously concerned about the economic future of the university.

The one highlight was that three weeks after 9/11, if you really dipsticked world opinion, the goodwill toward the United States was at a record high. So, five years later, where are we? That's really the topic of exploration for this panel. I think that there are many things we can look at, including predictions which were made that didn't happen, or at least didn't happen as permanently. Of course, many things did happen which were not predicted. Here and globally.

Let me start off by turning this over to Joe Petro.

JOE PETRO: Thank you, Dean Baczko. By way of introduction, I have worked for Citigroup for the last fourteen years; before that I spent twenty-three years as a Special Agent and Senior Executive of the United States Secret Service. Citigroup is a company with about three hundred thousand employees. We have in excess of 200 million customers and we operate in 105 countries around the world in about twelve thousand facilities. Last year we had over nine-thousand security related incidents. Of those, we characterized 193 as terrorist- related events. Ten were bombings, and eight were kidnapping of our employees. Here in New York City, we have a very large presence. We're the second largest employer in the city, with twenty-seven thousand employees and we have seventeen locations around the Five Boroughs.

Several years ago, Director of the FBI, Robert Mueller said, "there will be another terrorist attack and we'll not be able to stop it." That's a pretty pessimistic point of view, but he is, in many ways, correct. We are going to be able to stop a lot of them, but they only have to be right once and we have to be right all the time.

Since the Cold War, terrorism has changed the definition of the threat and the definition of the enemy. Most of us remember the Cuban Missile Crisis as the most dangerous period in our lifetime, in terms of how close we came to nuclear war. But, if you think back to that period, the adversaries were in the know and they were rational. No matter what you thought of Khrushchev, he was not an irrational person. And that crisis also had a time limit. It was thirteen days long. It had a beginning and an end. And ultimately that crisis was solved.

Today's terrorists are largely unknown. They could be in our midst. They are clearly irrational. And this crisis really has no end. It's something that we're going to face for the rest of our lifetimes. It has changed our protective tactics and our attitudes about security. Who thought a few weeks ago that we would not be able to take liquids or lipstick or things like that on airplanes? That certainly was not anticipated. What is unthinkable today may become very routine tomorrow. This is a reality that we all face.

I would like to make some quick, general observations. As many of you may know, eighty-five percent of the infrastructure in the United States is in the private sector. And there are many things in the private sector that we cannot do very well. One of them is protect ourselves against an organized terrorist attack.

We can't effectively anticipate or predict some of the new security risks they're developing and we can't depend on vague and imperfect intelligence information.

Terrorists control the agenda, the timing, and the location of the battlefield.

What are some of the things that we *can* do in the private sector? We can take some reasonable precautions. We can erect vehicle barriers; we can remove unnecessary signage; we can screen visitors; and we can move non-customer-facing businesses to low profile facilities. We can disburse key business functions; we can increase security guard presence; we can extend perimeters; we can perform some sort of surveillance detection; and we can conduct training programs. But there are some realities that exist, not just in the private sector but in all of our country, that we cannot prepare for.

One of these realities is that a sufficiently motivated attacker will always be able to outsmart a static defense.

This creates some additional realities for the private sector. First, absolute security is not possible in a corporate setting. Intelligence is imperfect, often wrong, and untimely. There are limited defensive capabilities available to the private sector. We place a tremendous dependence on static defenses. We have a worldwide exposure. Instead of prevention and protection, it's now response and recovery. Guards, gates, and guns are no longer enough. Finally, an urban environment has tremendous limitations, though that is another whole topic in itself.

One of the problems we face in New York and around the world is sustainability. I know the New York City Police Department has this problem. To keep the level of police presence high and sustain that for long periods of time is very difficult; it's equally difficult to do that in the private sector. We have to think about reducing the consequences of loss and balancing dispersal with concentration. Clearly, concentration is economically and efficiently better, but it's not necessarily good to put all your people and businesses in one facility. There must be some dispersal plan, particularly in an urban environment.

And there are significant employee issues. Not just after 9/11, but anytime there's a threat or change in the environment, we have to think about the employees and the emotional impact of these events. Our employees are looking for visual assurances. They want to feel safe. They may not necessarily be safe, but they need to feel safe. A lot of what has to be done is really for their benefit. There has to be good communication, but we also have to deal with the fact that we have to run a business and business has to continue.

Finally, there is complacency. I was at a briefing a few weeks ago in Washington and Director Mueller was there and I asked him what his biggest concern was for the country at the moment. He said complacency. I think that is a huge issue. It often takes some sort of event to get us back on track as time passes People have become more and more complacent about these issues.

Protection against terrorism must be a shared responsibility; we can no longer work in isolation. There's a new interdependency between the private and public sectors. We all really do worry about the same issues.

One of the biggest realities that we face in the private sector, particularly large multi-national companies, is that we're operating not just in big urban environments like New York, but around the world.

To sum it up, the private sector is vital, the private sector is at risk around the world, and the private sector must be involved in creating the solutions.

JOE BACZKO: Thank you, Joe, we'll come back to that in a minute, but now I'd like to bring up Michael Dolfman.

MICHAEL DOLFMAN: I'm from the U.S. Bureau of Labor Statistics. We're the number people and I always like to try to put the numbers I'm going to present in perspective. As part of my responsibilities as Regional Commissioner, I'm in charge of economic analysis for all the regional offices in the Bureau of Labor Statistics. Just recently, I was flying back from my San Francisco office and the plane was one of these new three engine fuel-efficient airplanes. We were about an hour out of San Francisco when the pilot comes on and says, "I've got some difficult news. We're having trouble with one of engines. It's nothing to worry about, but we're going to be about an hour late getting into LaGuardia." I looked at the fellow next to me, and he looked at me, and we didn't say anything.

Another hour later, the pilot comes back on and says, "I have more bad news, the second engine is giving us trouble. Now the only problem is that we're going to be about four hours late coming into LaGuardia."

"Well," the fellow next to me leaned over and said, "I hope that third engine doesn't go out; if it does we'll be up here all day."

The numbers I'm going to present come from two articles. The first one is "9/11 and the New York City Economy," published in the June, 2004, issue of the *Monthly Labor Review*. The next one is coming out in October of this year, where we're looking at structural changes in Manhattan's post 9/11 economy.

Manhattan is a very, very unique place, and its economy is very unique. We have 1.8 million private sector jobs; sixty-five percent of all New York City jobs are in Manhattan. We have an annual average wage in excess of $73,000 a year. We have the highest county wage—remember Manhattan is a county—in the entire country. And we're very unique. Manhattan is the only county in the United States that has more jobs than it has people—2.4 million jobs, 1.5 million people.

Since 1978, what we have seen in the United States is that employment has increased, but real wages, that's actual wages deflated for inflation, have remained relatively constant. In Manhattan during the same period, employment has remained relatively flat, but real wages have gone up significantly.

To understand the Manhattan economy, we have to understand globalization. About thirty years ago, the global economy, as we now know it, started to emerge. It was interesting; it was at a time, as some of you remember, that New York City was facing bankruptcy. People in the know tell me that for about four hours the city actually was bankrupt. But then something happened, the complexity of the global economy and the requirement for increased scale grew: that increased scale of transactions meant that there was a high need for specialized services,

36

including services in finance, marketing, accounting, computer, consulting. Manhattan was the one place that was able to provide all these services and as a result Manhattan became the capital of the global economy.

By the twenty-first century, Manhattan's economy had matured and bifurcated into two major components. One was the global economy; the other was the local economy. Both were related to each other, both needed each other, but both were vastly different.

In effect, 9/11 cost the Manhattan economy sixty-thousand jobs each month for four months, and 2.2 billion dollars in lost wages. Now, sixty-five percent of all the jobs that were lost in 9/11 and eighty-eight percent of the 2.2 billion dollars in wages were associated with the global economy. The global economy is what gives Manhattan its shape, its purpose, and its driving force. But 9/11 wasn't all that happened in Manhattan. The period from 2001 to 2004 saw a very serious economic downturn. As Dean Baczko pointed out, a lot happened at the beginning of the twenty-first century. We had the tech stock bubble burst on Wall Street. All the dot-coms went south. We had corporate scandals—Enron wasn't alone. We had an overall drop in consumer confidence in the country. And finally we had the Iraq war.

The economic decline from 2001 to 2004 was very unique because it caused a fundamental shift in the nature of Manhattan's economy.

During this period of decline, we had a loss of 177,745 private sector jobs in Manhattan. That's 9.4 percent of the job base. The global economy itself in Manhattan accounted for 4.9 billion dollars of lost wages and seventy percent of the jobs.

All of the 177,000 jobs that were lost in Manhattan were related to these industrial sectors: finance and insurance, professional and technical services, and the information sector.

Now, in a sociologic sense all jobs are equal. In an economic sense, though, all jobs are not equal. The jobs that were lost in Manhattan were the highest paying jobs. In finance and insurance, the average weekly wage, first quarter, was $6,861. Any students here? Change your major immediately to finance.

Professional and technical services—these are the lawyers and the accountants and the consultants and the computer experts—they had an average weekly wage of $1790. And the information sector, which includes publishing, movies, magazines, radio, and TV, averaged a weekly wage of over $2,000.

Within these sectors, the real loss took place in securities, brokerage and commercial banking, computer design, and publishing and motion picture.

One of the questions we're confronted with is how did the 2001-2004 economic downturn cause a restructure of Manhattan's economy? To answer that, we've taken a 15 year period, from 1990 to 2004, and looked at everything that took place then. And what we found is that the global economy had been the driving force.

There is a relationship, or had been a relationship, in Manhattan between jobs created in the global economy and jobs created in the local economy. From 1990-1993, as some of you will remember, there was a severe economic downturn in Manhattan. From 1993-2000 there was a tremendous expansion. Interestingly enough, using something called Economic Base Analysis—which we describe in the papers in much more detail than I'm going to do now—we found that during the downturn of 1990-1993, every job that was lost in the global economy resulted in a loss of 1.2 jobs in the local economy. These were people in the restaurant business, in arts and entertainment, in retail trade. During the expansion of 1993-2000, every job gained in the global economy resulted in a gain of 1.2 jobs in the local economy. The relationship held.

If we look at the latest turndown, however, we find something completely different. Job formation and job losses in the global economy no longer influenced anything that happened in the local economy. Instead of new jobs, it is high wages that today are associated with the global economy in Manhattan.

Just to give you a flavor of these high wages, consider the finance sector between 2001-2004, where weekly securities brokerage incomes went from over $10,000 to 11,000, a 9.6 percent increase. Portfolio management saw a 29.6 percent increase. Even though we have lost jobs, a significant number of jobs in our global economy, we have seen dramatic increases in average weekly wages in these industries.

In 1990, 47.3 percent of all Manhattan jobs were global economy jobs. In 2004, that dropped to 45 percent. But if we look at effective total wages we see that the wages people earned in Manhattan from global economy jobs increased from 66.4 percent to over 77 percent.

Fortunately, in 2005 we began to see the beginning of a turn around in Manhattan, an increase of about 1290 jobs, or a tenth of a percent. That could just be noise, it could just be statistical variation, or it could be an actual increase.

Interestingly, when we track the numbers regionally, we see that all the job growth in the New York Metropolitan area—from New Jersey to the northern suburbs to Long Island—took place in New York City. Interesting, too, is that we're seeing tremendous growth in the finance sector. Apparently all these back office jobs that moved out after 9/11 are coming back. However, unfortunately, in the city as a whole, we are still 150,000 jobs less than we were at the peak in December of 2000.

Thank you very much, I refer you again to those two articles which will give you a lot more information than I was able to present.

JOE BACZKO: Eric, do you want to make some comments on what's been said so far or go ahead on your own? You're with the Alliance for Downtown and very intimately involved with these discussions.

ERIC DEUTSCH: Thank you, Dean Baczko. I'm going to make a few brief comments and be a little bit more parochial in terms of our interests in Lower Manhattan.

As many of you know, the Alliance of Downtown New York is the business improvement district for Lower Manhattan—basically the area south of City Hall, south of where we are at Pace. Obviously, I'd like to thank President Caputo for putting this together. He is one of our board members at the Alliance and a great friend to the work that we do.

A couple things that I'd like to point out that we've been recognizing in the market in Lower Manhattan are things that have evolved over the last five years. Certainly after 9/11 there were a number predictions about what would happen downtown, what would happen in New York, how many jobs would be lost, how many companies would leave, how nobody would want to live down here again. Fortunately, most of those predictions proved false. Now, those predictions likely were not based in any scientific research but instead were based on people's feelings and conjecture. At the time, I was working at the city's Economic Development Corporation, which is a group involved in New York City government and economic development. We were a lot more optimistic, a lot more focused on what we could do for the recovery.

I'm pleased to say, as we've been saying for a while at the Alliance, that we've moved beyond the point of recovery and have begun to turn the corner to full-fledged growth in Lower Manhattan. Not long after 9/11, with the priming of the pump that is often done by the government in a good way, a number of incentives were provided for people taking apartments in Lower Manhattan. As a result, that market really began to take off as a function of the overall New York City market. In 2001, there were about 23,000 residents of Lower Manhattan, south of Chambers Street; today there are 37,000 estimated residents in Lower Manhattan, south of Chambers Street. There was such dramatic uptake in the residential market that the rental market was no longer viable for developers—only the condominium market was viable.

Residents are pioneers. You'll find this in neighborhoods all around New York City. The waterfront in Brooklyn, in Williamsburg, Redhook, the Lower East Side, Upper Manhattan, various places that over the years have changed. People didn't expect residents to come downtown, but they did. And residents will go where others typically won't. Businesses, retailers, usually don't lead the way, but they'll follow along afterwards. In the last couple of years, we've seen a dramatic uptick in the retail trade in Lower Manhattan. We estimate that there is about a billion and a half dollars of unmet demand in terms of retail trade in Lower Manhattan. That is the buying power of the residents and, more importantly, the 300,000 plus people who work downtown everyday. Now we've seen that the retail community *has* recognized that the buying power is there. There's no need to wait for the Trade Center to be rebuilt; there is no need to wait for the new owners of the South Street Seaport to reposition that property. Hermes, Tiffany's, Hicky Freeman, BMW, Sephora, and a variety of other independents already have taken spaces downtown. They see that the buying power is there. That's been a very exciting part of our turn-around and recovery in Lower Manhattan.

The other thing that we were very hopeful would pick up was the office market, the commercial office sector. That sector over the last year and a half has picked up dramatically. We track these numbers pretty regularly and from the beginning of 2005, 1.7 million square feet of tenancy and dozens of companies have moved to Lower Manhattan who were not here before. Some are consolidating down here, like Aon who had 400,000 feet in the Trade Center and moved half of it to mid-town and now look to be moving that half back downtown, which is very exciting. This really shows us that there's a lot more confidence in the long term prospects of Lower Manhattan. These business-es don't make a one-year commitment, like a resident signing a lease. They make a ten-year, fifteen-year, twenty-year commit-ment when they sign a lease.

We've also begun to see an uptick in rents, a decline in vacancy rates. Vacancy rates are edging below ten percent and rents pushing towards $40, ultimately to the point where we'll have a self-sustaining central business district where it's worth-while to make investment in office property down here.

The other thing that is very encouraging is the diversity of the economy in Lower Manhattan. We see that more and more of the businesses that are coming downtown are outside of the financial services core. We've seen publishing, non-profits, pro-fessional services, architects, engineers, and lawyers. There are all different types of educational institutions that are expanding and there are healthcare providers who are taking space down-town. That, to us, is very encouraging. In Wall Street terms, you want to diversify your portfolio. Well, we think Lower Manhattan has really begun to diversify its economic base. So when Wall Street has its ups and down, Lower Manhattan will feel them less than we used to. So that's been very encouraging over the last couple of years.

The other thing that we've noticed is the number of smaller businesses that have come downtown. The number of businesses in Lower Manhattan is now in excess of eight thousand. That's a number that we had not seen since prior to 9/11. And while we have fewer jobs downtown—fewer private sector jobs, because we have less inventory of space—we actually now have the same number of businesses. The average business size has dropped from about thirty-six people per company to about twenty-seven per company. The interesting thing to note is that, as many economists will tell you, more and more smaller businesses downtown, companies that take the five or ten thousand square foot spaces, is where the future job growth is. Today's five thousand square foot tenant is tomorrow's twenty-thousand square foot tenant who, ten years from now, may take those five floors at the Freedom Tower.

So it's very exciting for us to see the great diversity of the economic base in Lower Manhattan. We are now at the point where people can make investments in their office properties and expect to get viable returns. That's always the target for a central business district. It's something that we think is very important because today we have about 210,000 private sector jobs in Lower Manhattan; at our height in the late 90s we had about 260,000. But when you look at that as a ratio from the overall amount of inventory, considering we lost about twenty-five million square feet of office space since 9/11, we're at about the right ratio. Half of that space was lost in 9/11 the other half lost through conversion. But to the extent that we have the inventory, Lower Manhattan will house the jobs. The people do want to do business here. The outlook is bright for Lower Manhattan.

JOE BACZKO: Steve, you've been President of the Real Estate Board for 20 years. Can you comment on what Eric just said? Do you agree with him on the residential and commercial prospects that he played up?

STEVE SPINOLA: When you look at all of these statistics and numbers and you look at where we are today, only five years after 9/11, it's an amazing statement about this city. Before 9/11, we had a vacancy rate in Lower Manhattan of 4.5 percent; right after 9/11 that number increased to about twelve percent or thirteen percent. And there was another ten percent, at least, that was leased, but not occupied.

Right after 9/11, we had a gathering with Mayor Giuliani. One of the topics we discussed was, how do we replace the fifteen million square feet of office space that we just lost. And nobody in the room realized that we didn't have to replace the fifteen million square feet at that time. Because there was space that had never been occupied, but was being taken up because of the perception that the market was hot and everybody was projecting expansions.

Back then, after 9/11, everyone was predicting an end to tall buildings, an end to cities as we know them. As I said to the dean earlier, I kept telling the press, "Nobody's going to go into tall buildings. We don't need to build tall buildings back down at the World Trade Center." Well, my office is on the second floor of a 60 story building, and not one tenant from the top half of the building has come down to suggest that we trade our spaces. The fact of the matter is, people are still prepared to go into tall buildings; it's exciting, it's what makes this city.

As Eric mentioned, right after 9/11 the owners of most of the residential buildings in Battery Park City gathered in my office and talked about the fact that they basically had 50 percent occupancy as a result of September 11. It wasn't that there were 50 percent available leases or apartments. I mean, people walked away from their leases, people refused to come back to their existing apartments. This was a period of uncertainty because of what the conditions were, of whether there was going to be another attack, of whether New York was still a prime target.

We spent the next year trying to figure out how to attract tenants back. It's funny, but money does work. The state of New York offered $12,000 over two years to any tenant who signed a two year lease or bought an apartment in Lower Manhattan. Before the program was adopted, the residential brokers who were my members, told me that they spent the weekend showing people these apartments in Lower Manhattan. Within a couple of years we were back up to the occupancy rate in Battery Park City that was before.

And since then, as Eric said, the programs to encourage conversions of obsolete office buildings to residential buildings created something we have been dying for; it made Lower Manhattan a 24/7 community. What is happening in terms of commitment to the infrastructure, the decisions, and the investment that's taking place all point to the fact that Lower Manhattan has come back.

We now have a $20 approximate differential on Class A office buildings between Lower Manhattan and midtown Manhattan. That's one of the reasons why we're seeing tenants making the decision to come down to Lower Manhattan. When you look at trophy buildings—I shouldn't name trophy buildings because my members will be upset if I don't mention their buildings—but trophy buildings are going for, in some cases, $100 a foot in midtown.

And the trophy buildings in Lower Manhattan, which would be the World Financial Center and Seven World Trade Center, are basically making deals today in the $50-60 range. Just eight months ago, we were told that rents would be $35 a foot ten years from now, that the office market was really not going to recover in Lower Manhattan. But when midtown gets too hot, Lower Manhattan absorbs. And when Lower Manhattan gets too hot, Brooklyn all of a sudden creates some space, as does the great state of New Jersey.

So what has 9/11 proved in terms of how our economy can handle this constant fear that our first speaker talked about? This is clearly not going away and the last thing we want to do is to encourage people to forget about their safety or forget about the security measures. But it's demonstrated that this city is

44

desirable as a city, that the people recognize the city for what it's worth. As one of my members told me, where else can you go to about five meetings in one day without ever having to get into a vehicle? I mean you can basically walk this city, except between midtown and downtown, and you can get down here in about eight minutes on the Lexington Avenue line.

One of my residential brokers was asked why people are paying so much for apartments now, after 9/11. Her response was—it was Barbara Corcoran—that it was because the world saw how we responded, how we could take this devastating attack on our city and continue to move. And think of it, during the last five years, we have broken every record in terms of land prices, in terms of sales of condominiums and co-ops, and we're now starting to break those records in terms of rents.

As well, we have broken records in the number of housing permits issued for the last four years. This past year, we had thirty-one thousand residential units permitted. They won't all be built, but most will. Each year we're growing. There is a confidence in the city because despite what the Mayor of New Orleans might think, we've done a phenomenal job of dealing with one of the most complex structures ever destroyed. Now, we're moving forward with transportation commitments and, if that's not positive enough, we continue to be the financial capital of the world.

The financial industries that are so critical to Lower Manhattan are back into a hiring phase. In terms of the economy, we're in phenomenal shape. I am concerned about the overheated residential market. But in the last six months we haven't seen prices go up. We've seen transactions slow down dramatically, but prices seem to be somewhat holding, not increasing. And in terms of jobs, we're seeing dramatic interest in the office market in the city of New York and we're seeing rents . . . there are reports of some buildings in midtown renting at $175 a foot, which clearly sets records here in New York. For me, that's the major test of whether or not the private sector is making commitments to the city of New York. They're clearly doing that, and they're clearly doing that in Lower Manhattan, primarily because of the terrific response by our local city government, our state government, and our federal government to 9/11. So I'm pretty optimistic and I think we ought to look at this fifth anniversary and recognize how well we've done considering the terrible and devastating attack that it was.

JOE BACZKO: I'm conscious of the fact that we're running out of time, and I promised we'd turn this over to some questions from the audience. If you have a question, raise your hand, as you have gentlemen. Thank you.

AUDIENCE: What effect would a second attack on the World Trade Center have on the economy? Is there a financial contingency?

JOE PETRO: You know, I don't want to be the one that says something is going to happen again, but I think if you look at the intelligence information, if you look at what law enforcement is concerned about, if you look at the rhetoric, New York continues to be the target of choice. I think if they thought they could pull something off in New York they would. I think all of us who live here are concerned about that, but we have the best police department in the world. I have a lot of confidence that even though New York may continue to be a target the terrorists are going to have a tough time getting through the NYPD.

Having said that, we know the United States is still going to be the target. If you look at the rhetoric of Al Qaeda and some of the other groups, it's the U.S. economy they want to take down, and it's Americans they want to kill. Where else can you do that most effectively but in the United States? I mentioned in my earlier comments that complacency is a problem. The U.S. business community has to continue to operate; it's vitally important. These very optimistic predictions and reality of what happened in New York since 9/11 are heartwarming for all of us, but I think we have to continue to be vigilant; this is everybody's job. Obviously we depend on our police department to defend us, but I think companies are in a difficult position. You know Citigroup has not really moved anyone out of New York since

9/11. We have twenty-seven thousand people here. We have seventeen thousand people here below Canal Street. We have a huge presence here in Lower Manhattan, and are committed to staying here. I think many other companies feel the same way. New York is a defiant city, and I think we will defy these people and we will win.

JOE BACZKO: I think maybe the question is more appropriate for another panel, but I do think we live in a world of controlled anxiety right now, both commercially as well as individually. I think you very well pointed out, Joe, just how huge this security industry is. I think that is probably one of the consequences of 9/11. It is a major industry now compared to where it was. Here at Pace, you pass by one hundred guards just to walk into the campus. I think we are, as Joe said, not completely safe, and I don't think we ever will be until we address some of the other issues that are probably more important than just the security.

Gentlemen, thank you very much. It's been a very enlightening panel and I want to thank you for your time.

PANEL THREE

ENVIRONMENTAL CONSEQUENCES: PUBLIC HEALTH

JOHN CRONIN: Good afternoon everybody. Welcome to our third panel today, Environmental Consequences: Public Health. My name is John Cronin. I'm the director of the Pace Academy for the Environment here at Pace University.

It's a challenge to step back from the immediate events of September 11 and understand the full breadth of the environmental consequences and the health consequences of that day. I'm accustomed to approaching environmental issues from the point of view of an environmental impact statement. So, in preparation for today, I put together a list of information that would have appeared in an environmental impact statement of the attack, had it been issued before September 11. This is what it would have sounded like:

• There will be thousands of tons of pulverized asbestos, glass, heavy metals, and dioxin PCBs released in a cloud of uncontrolled emissions equivalent to dozens of asbestos factories, incinerators, crematoria, and a volcano.

• The dust settling out from the cloud in the immediate vicinity of Ground Zero will cover sixteen acres, more than six feet high.

• 1.2 billion tons of solid and hazardous waste will have to be disposed. The Freshkills landfill will have to be reopened, even though it's not engineered to accept hazardous waste.

• 32,000,000 gallons of contaminated water will be pumped from the pit of Ground Zero, through PATH tunnels, to the Hudson River.

• 4,000,000 gallons of toxic, contaminated runoff from fire fighting operations will be pumped into the Hudson River.

• 203,000 cubic yards of dredge materials containing thirty years of toxic substances will be dredged from the Hudson.

• Hundreds of tons of particulates and pulverized materials will fall into the surrounding waters turning airborne contaminants into waterborne contaminants and smothering long stretches of the Hudson River bottom.

• There will be extensive damage to cultural and historical resources and structures.

• There will be long term contamination of residences and public spaces including furniture, ventilation systems, and ambient dust.

• 40,000 emergency responders and 400,000 people working and living within a mile of Ground Zero will be exposed to record levels of airborne pollution and toxic contaminants.

• Seven percent of Ground Zero workers will suffer immediate respiratory illnesses and sixty percent long after.

• Up to half or more of rescue workers will develop respiratory system deterioration, equivalent to twelve years of aging.

• There will be thousands of cases of long term psychological trauma for survivors, responders, site workers, residents, and others.

• There will be an immediate loss of three thousand lives and additional unquantifiable premature deaths due to the consequences of the event.

I'm stating the obvious when I say such an environmental impact statement would never be accepted by regulatory authorities. But that's because these are unmitigable. It was an event of such immensity and such proportion that there is no way that you could mitigate them once they were launched. They happened.

Our visual icon for September 11 is the crumbling of the World Trade Center. There is, I think, a tendency for many of us to contain everything in those few seconds. In fact, what was launched on September 11, when we're talking about the environment, and environmental health especially, was a slow-motion catastrophe. The better icon would have been the cloud of dust that was sent up, which took a very long time to settle out. As you'll hear more today, that cloud has not settled out yet.

Yesterday, a study was released by Mount Sinai which reminds us that the long-term consequences, the long term environmental health consequences, are very much with us and raises some serious questions about what we will do about the long-term impacts of September 11.

Our panel today is going to talk about the issues of response to these long-term health consequences. In our discussion, I hope that we will have an opportunity to talk about the adequacy of resources and what kind of planning we can still do to deal with the short-term consequences of those immediate victims who have not yet been identified. With me today on the panel are Lorna Thorpe, Deputy Commissioner of New York City Department of Health, Mental Health, and Hygiene, Division of Epidemiology; Bruce Logan, President and CEO of New York Downtown Hospital; David M. Newman, Industrial Hygienist with the New York Committee for Occupational Safety and Health; and Congressman Jerry Nadler, who represents the eighth Congressional District in which we are all sitting.

LORNA THORPE: Thank you, it's a pleasure to be here.

Dr. Cronin just set the stage with the environmental impact statement. A very powerful statement. I think it is almost difficult to imagine that this is truly the extent of the exposure we all faced.

Beginning on September 11, hundreds of thousands of people, well over 300,000 individuals, were exposed to the dust and debris cloud created by the collapsing of the Twin Towers and the adjacent buildings.

Over time there came to be a great array of exposures for those involved in both the response and recovery work on the pile and in the adjacent zones. Community residents and office workers were exposed to smoke and fumes from the persistent fires in the pile for months on end. And certainly for office workers and for residents the exposure to indoor dust upon return to their spaces and re-occupancy was terrific. These are some of the major exposures. It is not an exhaustive list.

Here is a quick snapshot of what we do know, to date, from the medical literature. First of all, a number of studies that were established to track the long-term health of people exposed to the attacks have identified persistent respiratory symptoms among those exposed and higher symptoms in subsets that were particularly exposed to acute conditions, like being engulfed in the dust cloud.

The symptoms include sinus and nasal complications, post-nasal congestion, heartburn, hoarseness and throat irritation, shortness of breath, and wheezing, chronic cough. A number of different studies have demonstrated very high levels of respiratory symptoms—perhaps two-thirds of respondents, rescue and recovery workers, and community residents.

Clinicians treating World Trade Center-exposed individuals have identified a number of common conditions. Mainly, upper-area cough syndrome, asthma, and reflux disease. Studies that have come out this year, in particular those with large numbers of rescue and recovery workers, have shown persistent lung dysfunction. We have an accumulated body of evidence that shows symptoms are high. Indeed, respiratory symptoms were high one to two years after the event. Lung dysfunction was high one to two years after the event. It's remarkably frustrating to say what the exact picture is now. We don't have evidence of what level of persistence has occurred in respiratory symptoms five years after the event. This is a situation where the medical literature is lagging behind what people are experiencing.

Mental health studies have shown a similar picture of a range of elevated conditions. First and foremost, Post-Traumatic Stress Disorder (PTSD) has been documented to be elevated among a wide range of rescue and recovery workers, among community residents, and among tower survivors, as well. Most of these groups have shown an increased risk for depression and generalized anxiety and higher levels of substance use and abuse.

I thought I'd mention briefly some findings from the World Trade Center health registry, mainly because it's the registry that captures all groups that were exposed to the World Trade Center attacks. This was an effort by the New York City Health Department in collaboration with the CDC. It is, in terms of numbers of people being tracked over time, the largest effort in the history of the United States to track and understand the impact from the disaster. It was launched in September of 2003 and enrollment was completed in November of 2004. A total of

71,473 people took part in a thirty minute health interview using multiple languages. In November 2004, when the enrollments were closed, preliminary findings were shared with the public.

Here are just a few examples to give you a sense of the magnitude of exposures. More than fifty-five percent of enrollees reported witnessing at least one traumatic event. Forty-three percent were actually caught in the dust and debris cloud. Fourteen percent evacuated from a damaged or destroyed building. Twelve percent evacuated from a home in lower Manhattan. And this is just among those enrolled in the registry.

The preliminary health finding from the registry was that sixty-seven percent of respondents reported at least one new or worsening respiratory symptom. There were many respondents who had more than one respiratory condition. The most common symptoms were sinus problems and shortness of breath, followed by wheezing, throat irritation, and persistent cough. As many as fourteen percent of the enrollees—and fourteen percent of 71,000 is a large number—had five or more respiratory conditions that were new or worsening after 9/11.

In April of this year, we published a specific report on the physical and mental health of individuals who were survivors from damaged or destroyed buildings. These were mainly tower survivors or survivors from some of the buildings surrounding the towers. There were about 8,000 in that sample. As with the entire registry, symptoms of psychological distress were very high. But we found, consistently, that those who were caught in the dust cloud had much higher levels of symptoms and distress than those who were not caught in the dust cloud. That's something that each of the studies that has come out in the last couple of years has shown.

We have a paper that we will make available to the public soon that looks at Post- Traumatic Stress Disorder in Lower Manhattan residents. This was on a sample of eleven thousand residents. We screened for actual probable current post-traumatic stress disorder two to three years after 9/11 and the prevalence of PTSD was thirteen percent. That's more than twice what we would expect in the general population.

I'd just like to end with some challenges ahead. I think everyone sees that there are major concerns with the persistence of respiratory symptoms. Some of these symptoms may bode a risk for more serious lung disability. There already have been case reports of individuals with serious lung conditions. But we certainly don't have information on how many people are suffering from respiratory symptoms that are common today and we don't have information on how many people are suffering from serious respiratory conditions five years after the event. This is really important: the more information we can gather, the better we can diagnose and treat patients moving forward. So it's important to know exactly what is happening today.

Many of the health conditions that we're talking about are common and they're not always recognized as World Trade Center related. This is a challenge for clinicians. We also know that gaps [in treatment coverage] do exist and they are not very well defined. The Mayor announced yesterday an expansion of some services for those who were not eligible for the first responder programs. Hopefully this will cover some of the gaps that have existed to date. But I think there's a lot more work to be done.

Thank you.

JOHN CRONIN: Dr. Bruce Logan.

BRUCE LOGAN: It's a great privilege to be here today. I'm the President and Chief Executive Officer of New York Downtown Hospital. As you know, we're the only hospital, really, in the Lower Manhattan area. We get thirty-three thousand emergency room visits per year, fifty-six hundred ambulance deliveries.

On 9/11, we were obviously the first hospital to get casualties and within the first two hours we had 375 patients that were brought to the emergency room. Many of these early patients were very seriously injured. We had twenty-four patients that were admitted, nine who required emergency surgery and twenty-four who were transferred to other hospitals. Those patients were mostly terrible burn patients. For those of you who have seen these kinds of patients, you know they are probably the most heartbreaking type to take care of. We sent a number of those patients to the Burn Unit up at Cornell. We had three patients who were dead on arrival and two subsequent deaths and we treated hundreds and hundreds of unregistered patients.

We also sent twenty-two physicians to Ground Zero to assist in the original triage stations because we expected, at first, that patients would continue to be brought to the hospital. Over the next week, we provided emergency medical treatment to over two hundred injured firefighters, police, and rescue workers, and provided thousands of meals to rescue workers, volunteers, victims, and local residents. We had an electrical generator so we were able to prepare meals.

We began to act as not only a hospital, but also as a source of refuge and food and water. We also delivered medications and hot meals, particularly to the South Bridge Towers. We had very many volunteers there and we have always had great communications with South Bridge, so they let us know who needed food and water and medications. It was sort of a partnership, in a sense, between the hospital and the neighborhood when responding to the issues of the many people who were displaced during this period of time.

Right after the plane crashed into the North Tower, the hospital declared an emergency. Very quickly, by 9:30 A.M., we had so many patients that we had to clear out the cafeteria and turn it into another area to see emergency patients.

Shortly after the collapse of the first tower, we set up the small decontamination facility that we had at the hospital at that time. We had patients coming to the hospital who were covered in dust and debris. And, then, of course, the next tower fell and the situation got even worse.

At the time, you would see people with respirator masks, but they were not wearing the mask. Later you might see someone who was wearing a mask and other people who didn't even have masks. The masks that people did have, whether they were wearing them or not, were all different. What occurred to me was, if you can see the air and it's not misty water, then it isn't good to breathe.

It was obvious, quite honestly, that there could be long term problems from breathing in air that was filled with pulverized buildings, possibly asbestos, dioxins, and the other types of poisonous materials that were in these buildings. It should be no surprise that it would cause problems. At the time, on 9/11 and in the following days, we did see a lot of people coming in with a cough and with asthma attacks and so forth. And over a subsequent period of time, we began to see what's come to be called the World Trade Center cough—a lot of coughing and clearing of throat and the other symptoms that were so well described by our previous speaker.

What we don't know yet is what the breathing in of these chemicals is going to mean to people in terms of cancer of the lung, cancer of the bladder, or cancer of something else. This is why the Mount Sinai registry and the Mount Sinai study and the World Trade Center registry are so important—to follow these people to see what the long-term health consequences of 9/11 are.

We already know the short-term consequences: When you breathe in the stuff, it's an irritant. People cough and have asthma attacks. And on a long-term basis we know there are decrements in the lung function of the people who have been tested. But we still do not know what the true long-term consequences will be.

This is more or less a repeat of what you had said earlier, that exposures to particulate matter and air pollution, such as soot, can have severe health effects. Theoretically, you could get heart attacks, asthma exacerbations, and so forth. And people who were already ill or people who were infants and the young are the ones with the most at risk. I'm going to end, because I want to leave time for our other speakers, but will say that as I look back at what happened shortly after 9/11, it is quite amazing that there wasn't a specific directive for people working at the site, to wear the appropriate protection and masks.

JOHN CRONIN: Thank you, Bruce. David Newman?

DAVID NEWMAN: Thanks for your interest and concern in this topic.

The events of 9/11 and thereafter are arguably New York City's worst environmental disaster ever. And, at the same time, they constitute one of the most significant failures of public health policy in our nation's history. This is what this issue is really about. Environmental health is about public policy, public health policy. In terms of environmental health, the broad nature of the response on and after 9/11 was shaped by political considerations, rather than by concern for the protection of public health.

It was characterized by a failure to enforce applicable environmental and occupational health standards such as OSHA's respiratory protection standards. The failure of the response highlights the inadequacies in the regulatory framework in dealing with natural or technological disasters such as this one. It subjected responders to significant health risks which were largely avoidable and illegal and from which they are now ill or in some cases are dying. It unnecessarily subjected unknown numbers of other workers, residents, and students to risks yet to be determined.

And last but not least, our failed 9/11 response has become official government policy and was largely repeated on the Gulf Coast in the aftermath of Katrina.

We're five years down the road from the events of 9/11 and we still do not have any good understanding of the nature or extent of the contamination that was associated with these events. Government agencies have taken thousands of environmental samples, but no agency has implemented a systematic, comprehensive indoor testing program or has attempted to collect and analyze the extensive sampling data that remain in private hands.

As a result, even at this late date, five years later, it is impossible to characterize risk or safety in any scientifically valid manner. As recently as last year, EPA issued a report entitled "Health Effect of World Trade Center Collapse." In that report, EPA stated that "short-term health effects dissipated for most once the fires were put out. There is little concern about any long-term health effects." That report's still on the web.

Unfortunately, as we all know, there is considerable evidence to the contrary. It's well documented that increasing numbers of rescue and recovery workers are suffering catastrophic illnesses and even fatalities. There is more limited research and anecdotal evidence that also indicate increased rates of illness among residents, though to a lesser degree. We must ask, are these illnesses that we see today the leading edge of the wedge? That is, will additional responders become sick? Will additional downtown residents and workers suffer adverse health outcomes as well? Note that I'm raising these as questions not as assertions. However, they're questions for which we do not yet have the answers. When we talk about risk and we talk about the possibility of adverse health impacts we need to consider the potential for exposure, because that's where the risks originate.

In terms of risk and in terms of exposure to the environmental fall-out of 9/11, there are distinct exposure populations. Not every individual or group that was exposed had the potential for identical exposure. I'm going to run down a number of exposure populations. Each of these populations, because we're talking in generalities, has a different potential for risk and for type of exposure.

The first exposure population is those persons caught in the dust cloud on 9/11. The second exposure population is the workers and volunteers at Ground Zero and at Staten Island and those involved in the waste transfer operation. The third exposure population is workers, including janitorial staff and immigrant day laborers, who engaged in the regular clean-up of the World Trade Center dust and debris outside Ground Zero. The fourth exposure population is workers who engaged in the restoration of essential services in Lower Manhattan, such as telecommunications, sanitation, water, etc. The fifth exposure population is workers who are currently engaged in the demolition of heavily 9/11-contaminated buildings in Lower Manhattan, such as Deutsche Bank at 130 Liberty Street, and

Fitterman Hall at 30 West Broadway, and others. The sixth exposure population is residents, other workers, and students who remained in or returned to contaminated or potentially contaminated indoor spaces in Lower Manhattan or in other geographic areas that may have been impacted. This population was potentially exposed and in some cases may still be exposed to 9/11-derived contaminants that remained indoors.

Clearly, we expect that the exposures and the resulting health impacts of those caught in the dust cloud and of those working on the pile were and are much more significant and much more severe than those of most other workers, residents, and students. But that does not mean that these other exposure populations were not exposed or are not still exposed or that there will be no health consequences.

There have been, and in some cases there still are, a number of external or outdoor sources of 9/11-derived contaminates that can infiltrate indoors. These include the dust cloud on 9/11, and the airborne combustion byproducts—the plume. As you know, that plume crossed over Brooklyn and New Jersey.

Particulates were disturbed and made airborne by rescue, recovery, and debris transfer operations at Ground Zero and along West Street and at Pier 25, the site of the barge operation. Recent, current, and upcoming demolitions, and future reconstruction activities at Ground Zero, which have the potential to go on for the next ten years or longer, have the potential to disturb and make airborne 9/11-contaminated particulates.

I'm going to finish by saying that this type of response constitutes a massive failure of public health policy. First and foremost, we have to acknowledge and address the illnesses that are coming to the floor now and which may continue to come to the floor in the future. This includes a need for not just recognition, but for adequate funding, staffing, and resources for access to expert healthcare for all impacted individuals and communities.

Second, although I think it's unlikely that there remain widespread residual contaminations at a high level of concern across broad slots of downtown indoor spaces, it is certainly possible that some indoor spaces remain contaminated. These, most likely, are closer to Ground Zero and most likely in areas such as mechanical ventilation systems. Therefore, five years later, we still need to identify any remaining indoor contamination, to assess it, and to abate or remediate it if warranted.

And third, briefly, what are the lessons learned? There have been a lot of lessons learned, but not much has been implemented. We need to erect new buildings differently so that we have toxic use reduction. We need to strengthen our regulatory standards and we need to enforce them. We need reforms in emergency preparation. We need to redefine the first responder population to broaden it. We need to implement training, predeployment, rather than just in time or later. We need to redesign respirators so that they're useable.

And finally—and this is a very abbreviated list—we need to characterize the site of the disaster prior to allowing residents, students, or workers to enter the site. You cannot provide adequate protection unless you know what you're exposed to. So in the rush to judgment, the first couple of days aside, for the next ten months, it was inappropriate for workers to be working in that environment, both at Ground Zero and throughout Lower Manhattan, with the potential for exposure to toxic substances, without having identified those substances, and properly trained, equipped, and protected those workers and those residents who were told to clean up their own apartments.

Thank you.

JOHN CRONIN: Thank you, David. Congressman Jerry Nadler?

CONGRESSMAN NADLER: Thank you very much. The environmental and health effects experience after 9/11 constitute, in my view, a huge betrayal. A betrayal of the workers on the part of every level of government, federal, state, and city. President Bush came and said, "We will remember you." He remembered about a day and a half. It was a betrayal of the workers who worked on the rescue and recovery. A betrayal of the people who live and work in Lower Manhattan, part of Brooklyn, maybe even Jersey City, wherever the contaminants went—and we still don't know where that is.

Let me give you a little chronology. September 11 was a Tuesday. On Friday, as the Congressman from the area, I formed the Ground Zero Task Force to do various different things. One of the things we did was to retain immediately an environmental testing firm to do spot checks on a number of indoor spaces. Though EPA and others were doing outdoor spaces, we didn't see anybody testing indoor spaces, so we tested a number of apartments. Not many, five or six. And we found high levels of contamination in *all* of them, though we found that they were different. In one, you might have a high level of asbestos and a low level of mercury, and the next one would be just the opposite . . . so, very variated.

Within two days or three days after the disaster, Christie Todd-Whitman, the head of EPA, came out and, lying through her teeth and knowing that she was doing so, said that the air was safe to breathe and the water was safe to drink.

The Department of Health of the City of New York put on its website, as I think David mentioned, and it may still be there for all I know, it was certainly there for two years, "If you return to your home and you see World Trade Center dust, clean it up with a wet mop and a wet rag"—guaranteeing that anybody who did so would inhale some of the stuff, guaranteeing that unless the person who did so was trained and certified and licensed to do it, they wouldn't do it properly and stuff would remain in the porous wood surfaces and the drapes and the carpets, behind the refrigerator and the HVAC system, and so forth, to be inhaled in the future.

Then we had the rescue and recovery. For about fifty or sixty days, up to 40,000 people worked on the piles or at Fresh Kills. A deliberate decision was made not to enforce the OSHA standards. The people who cleared up the Pentagon wore respiratory equipment. The people who cleared up here . . . some did, some didn't. We had testimony that for many it was unavailable. That was a terrible dereliction of duty. I will make an exception for the first two or three days, when there was suspicion there might still be people alive there. We cut a lot of corners to try to get them out fast. But once we knew that there was no one else alive, it was simply a question of clean up and

61

recovery. There's no excuse for disobeying the law and letting people work in conditions guaranteed to produce sicknesses and early deaths. Frankly, people who did that, knowingly did that, should have been prosecuted criminally.

There was then a decision made, without properly securing the site, without properly classifying the site, to urge people to return to their homes and their workplaces, to re-open the schools. We know there was political pressure. Economically, Wall Street had to open its center. People returned to their homes, their workplaces, and the schools—in very many cases to contaminated homes, workplaces, and schools. We know nature cleans up the outdoors. The wind washes away the pollutants, blows it away, the rain washes it away. I don't know of any basis for assuming that indoor spaces get cleaned up at all unless someone does it over any length of time.

At my request, the EPA Ombudsman office held two public hearings in February and March of 2002. By the way, we publicly released the results of our findings in October, October 14th I think it was, of 2001. We gave the results to the EPA and to the newspapers; they were ignored. In February and March of 2002, the Ombudsman held hearings over which he and I jointly presided. One was held down here at Pace and one was at the Federal Courthouse, as I recall.

In any event, the EPA instructed other agencies of government to ignore the hearings and then put out a press release before they occurred saying that they were going to be a political charade—before they knew what was going to happen. There was testimony given by police officers that they couldn't get respirators. There was testimony that to properly inspect an apartment to see whether or not it was contaminated would be about twelve hundred dollars and that the cost of properly cleaning up a contaminated apartment would be, depending on the size of the apartment, ten to twenty thousand dollars per apartment to do it properly.

My office published a white paper in March and a revised version in April [2002] asserting that the duty of the EPA under the law was to lead, to come in and clean up these spaces, to do inspections and cleanup. We even told them the methodology

62

that they should use. And we were ignored. They then, under pressure, staged what I call a phony clean-up in the spring of 2002, starting in June. Phony because number one, they said they would only do it below Canal Street. As if there was a thirty-thousand foot high wall at Canal Street and the plume never crossed it; of course, we know it crossed it. As if there were no contamination north of Canal Street or on the other side of the East River; although people in Borough Park and Park Slope were picking up pieces of paper from the World Trade Center from their backyards. We know there was pollution all over the place.

Nonetheless, they geographically limited their intentions. They said they would only inspect apartments upon request. Well, I didn't request that my apartment be inspected. So maybe your apartment *was* cleaned up, but it would be re-contaminated from my apartment through the HVAC system. You have to do a thorough, building-by-building clean-up, not a partial clean-up.

Three years ago exactly, in August of 2003, the Ombudsman handed out a report saying that this was all terrible and as a result the Ombudsman's office was dismantled and no longer exists.

In August of 2003, the Inspector General of the EPA came out with a devastating report that called the clean-up a phony job. What they said had to be done essentially echoed the white paper that we had published a year earlier. He said, "What you have to do is to do inspections. Sample inspections in indoor spaces, in concentric circles out from the World Trade Center, as you might find that the problem goes for three blocks in one direction and three miles in another direction. And where you find contamination you must inspect every apartment, every workspace in that area, and thoroughly clean it up." None of that was done. A year later, under pressure from Hillary Clinton, the EPA set up a scientific panel and they started saying, "Gee, everything wasn't done right."

The Workers Comp Board in New York has controverted workers' claims at about five times the normal rate of controverting claims. The report that came out yesterday, that looked at 9900, (of the forty-thousand) workers who worked on the pile, found a seventy percent rate of sickness. We know that there were seventy thousand people who were either caught in the cloud or worked on the pile. Project that out, that's almost fifty thousand people. And that's not counting the people who are being slowly poisoned in contaminated apartments and work-spaces to this day.

What we need to do is three things. One, we have to pro-vide—and they said this at Mount Sinai yesterday—lifetime monitoring for all people caught in the cloud. All people. Currently, we have funding for monitoring for about another two years.

Secondly, we have to provide medical services. The Mayor announced the sixteen million dollar program yesterday. He's off by probably about two zeros. By two orders of magnitude in terms of what it'll probably cost, but it's a nice, first, baby step. And we have to have that proper clean-up, because, as I said, the internal air is not clean. We know that.

I usually say in my speech that I divided the victims into four groups. One of those groups is the people who live and work in contaminated environments. We don't know how many there are, we don't know who they are, but we know they're there. We also know that some people who get a small but chronic expo-sure to carcinogens are going to develop life threatening illness-es eventually. Some of them. How many, we don't know.

We also know it's cumulative. Even if we wave the magic wand today and clean up everything we know that a certain number of people, we don't know how many, are walking around doomed to develop lung cancer, asbestoseis and mesothelioma, and other diseases because of our malfeasance of the last five years. But if we don't wave the magic wand and we continue as we've done, then a much larger number of people will come down with these life shortening illnesses, fifteen or twenty years from now.

It is nothing short of immoral not to do this. In summary, we have betrayed—all levels of government have betrayed—the people who worked on the pile, who gave of themselves, not thinking of themselves, and who are now paying a heavy price.

We have betrayed the people who live and work, not only in southern Manhattan but in parts of Brooklyn, Queens, and Jersey City. We don't know how many because we haven't done that analysis. We have to reverse this betrayal. And the Mayor, instead of saying, "well, we're going to do a little," he should be leading the charge to get the Federal Government involved. Some of us have been talking about that for a long time— Senator Clinton, myself, Congressman Maloney—trying to get the Federal Government to stand up and do what it should, which is to pay for the medical costs, clean up the site, and learn from the lesson for the future.

JOHN CRONIN: Thank you, Congressman Nadler.

We are going to run out of time in a little bit and so what I want to do is turn it over immediately to the audience. What I would encourage you to do is ask some questions that look forward. Because I think if you listen carefully, each one of our speakers, each one of them, has raised some very serious questions about what the future holds for the living victims, what we're going to learn from events like this for the next catastrophe, and who's falling through the cracks.

AUDIENCE: Given that the particulate matters discussed were on the order of tens of microns, would a respiratory system with a filter be effective? Or, would we need an entirely self-contained sort of apparatus?

DAVID NEWMAN: I take your question really to mean that when respiratory protection is required, what type of respiratory protection would be appropriate in a situation like this. The people who were most exposed, or potentially most exposed, obviously were the people in the dust cloud and the people who worked on the pile at Ground Zero.

What was appropriate and what would have been appropriate under those circumstances? There are two basic categories of respirators. One is a supplied-air respirator, which you're referring to. One type of supplied-air respirator is like that when you get a canister on your back, like the firefighters wear, a Scott pack. The other category of respirators is air-purifying respirators, where you're breathing in the ambient air, but it's passed through a filter and presumably you want the right type of filter to allow the air through and exclude the contaminants from getting in.

At Ground Zero, the appropriate air-purifying respirators, N100s—that refers to the particulate efficiency of the filtration, which would have been equipped with organic vapor canisters; so dual canisters for the vapors, fumes, etc.—would have been adequate. But I would caution anybody to think that you can go to a store or go to the internet or use whatever source you have to obtain a respirator that will provide adequate protection. Because appropriately using a respirator is a fairly complex task that requires medical evaluation, requires fit testing, and requires a certain amount of technical training. In an employment situation, all of those things are required by law, and for good reason.

AUDIENCE: I was at the Trade Center on 9/11. I lived across the street on Albany and South Bend and developed health problems and so on from being there which I couldn't shake, even though our building had been cleaned, and my own apartment cleaned several times.

I moved to one of the Green Buildings in 2004 and the asthma cleared up. So the contamination was in my building. But the drain pipe that goes from the rooftop to the ground on the building on Albany Street has been completely torn out because it corroded from the dust, from 9/11. I mean, if that's happening to pipes you can assume that what's happening to our lungs is even worse.

Also, that building wasn't renovated until 2005. Now, five years later, they're renovating these residential buildings, ripping out carpeting and everything else, just as they would renovate any building, without taking any precautionary measures. These pipes that they just ripped out and threw away should've been submitted for some sort of testing because I'm sure it's going on in the other buildings. All the carpeting that was torn out, even the air quality in the hallways, nobody even thought to test any of that.

I mean it doesn't require the EPA to put a special program in place. This is all going on in the course of daily life. We just need to make people aware of it. They should report it or let someone know that it is going on in their building. It's happening now; now's the time to take advantage of it, because we're renovating. There's just so much that just happens from day to day, it doesn't require an act of Congress. That's my question: Is anything being done?

DAVID NEWMAN: In response to your well-stated concerns, in the absence of government policy, which is what we have here, we essentially have the Wild West, where anything goes. Like the Wild West, groups of tenants, groups of workers, groups of unions, and other groups have banded together and engaged in activities to attempt to protect themselves.

There have been circumstances, for example at Independence Plaza North, where tens of thousands of windows were to be replaced. Well, the Independence Plaza North tenants got together and negotiated with building management and were able to get all the windows and window wells tested and as a result of those tests were able to redesign the replacement process so that it would be more protective of tenants. There have been other cases, in other buildings, where tenants and tenant organizations have negotiated with the landlord over the process by which, for example, hall carpeting was to be removed.

Again, what we're concerned about is the absence of government policy. But even in the absence of government policy, vigilance is important and activism is important.

JOHN CRONIN: On behalf of Pace University and President David Caputo, I want to express our appreciation to Lorna Thorpe, Bruce Logan, David Newman, and the Honorable Jerry Nadler for being on this terrific panel. I want to thank you for being in the audience today, and join me in thanking our panel members please.

Be safe going home. Thank you very much.

PANEL FOUR

IF YOU LIVED HERE

JIM CAVANAUGH: Good morning everyone, and welcome to the panel, "If You Lived Here." My name is Jim Cavanaugh, and I'm currently the president of the Battery Park City Authority. We've got a very impressive group with us today. Tom Healy is with the Lower Manhattan Cultural Council. Certainly, as Lower Manhattan continues to recover and become a true twenty-four hour community, which is something that it's never been before, culture becomes an incredibly important part of that process, and Tom's done some great things. Charles Lai is the director of the Museum of Chinese in the Americas, and has done a lot of work in collecting information on the effects of the attack on Chinatown. Julie Menin is the chair of Community Board One, and before that was a founder of Wall Street Rising. And Mark Schaming is the curator of the New York State Museum and is responsible for the exhibit here at Pace which displays artifacts from Fresh Kills landfill. I'd encourage everyone to see it. We also have a much larger permanent exhibit relating to 9/11 in the State Museum up in Albany.

I did not live here or work here on September 11. I lived in Westchester, and I came to work in Battery Park City about two years after the attack of 9/11. Since then, I've been responsible for or able to witness much of the ongoing process of recovery. Battery Park City, for those of you who don't know, is ninety-two acres of former landfill just across from the World Trade Center site. In fact, some of the landfill from the World Trade Center was used to help create the foundation of Battery Park City. This neighborhood was developed according to a master plan, and certainly on September 10, the day before the attack, that master plan was reaching its height. Battery Park City had been subject to the vagaries of New York City's economy. When the economy was good, there was a lot of construction, both residential and commercial. When the economy wasn't so good, it slowed down. But we were very much nearing completion. On

69

September 10, 2001, we had residential construction underway and were two to four years from completing that master-plan; there literally was building going on on every available site. The question, of course, after September 11 was, "what would the future hold?"

The answer, as to what has happened in Battery Park City since September 11, is that it's a lot better than anyone reasonably could have expected in the days after the attack. I think if you had gone to anyone after that attack and said, "five years from now, here's what will have happened to Battery Park City," I think probably there would have been a great deal of skepticism.

Following September 11, Battery Park City had to be emptied out, temporarily. In fact, one of our larger buildings, Gateway, was vacated for a year. Since then, Battery Park City has become fully occupied. We literally have a zero vacancy rate or close to it for our residential properties, and the commercial properties are doing well. We just learned from Brookfield that they are renting space for fifty dollars a square foot, which is very, very good, and a very significant increase over the past couple of years. So, residentially and commercially, Battery Park City is truly back.

Between now and 2009, we're going to be building over thirteen hundred residential units at Battery Park City. We have deals with developers, so it's not just something we hope to do, it's something that's going to happen. And in fact, since 9/11, we've opened close to three thousand units. So, you're talking about approximately five thousand units of additional residential space in Battery Park City since 9/11. I think that's an incredible story of recovery and progress, and again something that was not in people's expectations right after the attack. And the really interesting thing—I hope Julie will talk to this, because she is a parent as well as a community activist—is that we are seeing a change in the types of apartments being built. Developers who came to Battery Park City before 9/11 typically built a lot of one-bedrooms and studios with the expectation it would be very much a singles, or small family community. What the developers are telling us now is how they've been surprised by the demand for larger apartments for families and people raising children. And to me that really speaks volumes about people's attitude down here, and speaks volumes about what the future holds. It's something that we find very, very encouraging. So, again, that change in the demographics is a good thing, and something that people did not expect in those very dark days after September 11. So what I will do is turn to my fellow panelists, and ask them to give you a little longer introduction of themselves and perhaps talk for five or six minutes about what they're doing, and what they see as the future. Tom, I'll start with you.

TOM HEALY: Jim Cavanaugh obviously comes from a much more modest and polite Irish-American family upbringing than I do, because he didn't really sing the praises that are due Battery Park City on so many fronts and I'd like to talk just about the cultural front. Battery Park City Authority is without a doubt, in terms of public agencies around the United States that deal with housing and commercial development, probably the most progressive and forward-thinking public agency in its commitment to the arts and culture. If you have a chance to go over to Battery Park City, you will see monuments, you will see public artworks. You will see some of the most innovative and beautiful gardens and parklands, and that is because of Jim and his colleagues. And if you are in Battery Park throughout the year you will also see performances of all kinds. It is really a model of integration of what we hope the whole of Lower Manhattan will become. So, he won't toot the horn of Battery Park City in that way, but I would like to. We at the Lower Manhattan Cultural Council are thrilled to be partners with them.

Just one little story about how Battery Park City works. I was going for a run there about three weeks ago, a huff and puff jog. It was a Saturday, all sorts of people out, Lady Liberty looking back at everyone, and there was a person reciting Walt Whitman's poetry to himself, you know, a typical New York City character. I noticed three people who were obviously just residents of Battery Park City, not people who worked there or who were gardening there, and they stopped actually to listen to this

71

person reciting Walt Whitman. It was a kind of quintessentially perfect New York moment, and it reminded me, when we think about the recovery from tragedy and what Battery Park City and the rest of Lower Manhattan were like right after 9/11, it reminded me of this crazy scene at the end of *Candide* where Pangloss and whole crew are going around the world trying to find this safe and crazy place. A decent place. The claim is that we should cultivate our own garden, that that's the hope that we can achieve for some kind of safety and calm and peace in the world. And that morning, that's just what I saw: the cultivation of a garden and someone reading poetry, and I thought well you know that's exactly what we'd like for Lower Manhattan.

The Lower Manhattan Cultural Council was actually founded thirty-three years ago by David Rockefeller, when the World Trade Center was first being built. The idea was to stimulate some cultural life in Lower Manhattan. As Jim was saying, Battery Park City at that point was just a landfill and a beach, and some of the early cultural activity that LMCC was involved with was staging cultural exhibitions on that beach called "Art at the Beach." Since then LMCC has grown into a presenter and supporter of a whole range of cultural activities. The heart of LMCC life was based at the World Trade Center. On 9/11, we lost everything: our home and our performance and exhibition venues. The residency we have for artists, too, was lost. So we, as an organization, spent a couple of years in the desert as LMCC was rebuilt. We refer to ourselves as a thirty-three year old startup. The last two years have been a dramatic and new time for us. We've been engaged in a whole range of cultural activities. And now, let me turn back to Jim.

JIM CAVANAUGH: Tom, thank you. You know, one of the communities that was hard-hit, perhaps more than most, after 9/11 was Chinatown. Chinatown was particularly devastated economically. It is located a little further from Ground Zero than some neighborhoods, yet because it is a neighborhood of immigrants, people perhaps were a little more economically fragile and had perhaps fewer reserves. I think there was a focus, a particular focus, on how that community would respond and whether it would thrive, whether it would continue the incredible growth that it had seen up until that point. Charles Lai, who is director of the Museum of Chinese in the Americas has put together, in the past five years, some exhibitions that documented the effect of the attack on the community, and I think we'd like to hear from him at this point.

CHARLES LAI: Thank you, Jim. These last five days, in fact this time every year, is quite an emotional time for me. I was on Wall Street, in the subways, just before the first plane hit. When I got out of the train station, the second plane hit. I spent the rest of the morning watching the flames and the collapse and people walking across the Brooklyn Bridge, making their way home. From that moment, there was the Before and the After. But even then, from the moment of the attacks and all during the week when I was glued to the tube, replaying those images, I felt that it was important to think about what else was happening. My family, my parents, lived in Chinatown; my sister worked in the World Financial Center; my nephew worked at the Marriott, and while I knew that everyone was safe, I also knew the impact would continue. Chinatown had been closed down; the economy in that area was frozen. I felt that it was important for me to spend time working to make some noise, to make certain that in the course of providing security for Lower Manhattan that our community would not be so disrupted that it would be irreparable. I spent a tremendous amount of time working to bring in the resources to deal with the multiple needs of the community.

After about three years of that work, I went back to the museum, of which I was one of two co-founders. Tom spoke in terms of the LMCC being a thirty-three year old startup. We became a twenty-five or twenty-four year old startup again. We felt we needed to bring a sense of life back to the community. Yes, the attacks of 9/11 were horrible, but our community is not inherently a place to sleep or to work. The life of Chinatown is determined by how we interact with each other, our sense of awe and culture, and our liveliness. So, in the last several years, the museum ventured into documenting some of the effects of 9/11 on our community.

JIM CAVANAUGH: Julie Menin heads the local community board and is raising a family down here. Before that she was involved in a startup, Wall Street Rising. She brings her perspective as a resident and as a community activist for the past five years.

JULIE MENIN: It's a real pleasure to be here today. I lived downtown on 9/11. As Jim mentioned, I still live here now, raising three children under three years of age. I really can't think of a better community in this city to live. We have truly pulled together since September 11.

September 11 affected me personally very much in that it caused me to change my career path. I was a corporate and regulatory lawyer, and owned a restaurant called Vine in the Financial District. After what happened after 9/11, I decided that I really wanted to be involved in the revitalization of this community, and I founded a non-profit organization called Wall Street Rising, which now has over thirty thousand members, and has been involved in a number of revitalization projects downtown. I chair Community Board One whose geographic area is south of Canal Street. As Jim mentioned, one of the key issues that our community is facing now is residential growth. Several months after 9/11, many people were extremely concerned about the residential occupancy rates, which had fallen from ninety-four percent to close to sixty percent. People were worried that we were not going to see residents stay in the community or move into the community. As a result, the Lower Manhattan Development Corporation implemented a residential grant program which really did lead to a massive residential explosion in our neighborhood, along with other factors that encouraged developers to come in and develop in our neighborhood. While the residential growth has certainly been good news, it has also meant that the schools are now severely overcrowded. P.S. 234, in Tribeca, is at a hundred and twenty-two percent capacity. So, as a community board, our top priority has been to find new locations for schools downtown. At the same time we're also very focused on finding increased recreational facilities for our youth, and, something I know that is near and dear to Tom Healy's heart, we're very focused on the Performing Arts Center at the World Trade Center Site, which we feel has been put very much on the backburner. If we could build a grand performing arts center, along the lines of a Lincoln Center, it would serve as a magnet for this whole neighborhood. It also would serve to attract office tenants, which is vital, given that we have a double-digit office vacancy rate downtown. So with that, I will turn it back over to Jim.

JIM CAVANAUGH: Mark Schaming is the director and curator of the New York State Museum. Though he doesn't live here in New York, he spent almost forty days combing through the remains of the World Trade Center at Fresh Kills to find suitable objects that could be used to create the exhibit which Pace is now hosting. So, while he doesn't live here, he probably has a greater connection to what happened on 9/11 than many who do, having looked at thousands upon thousands of relics and connections, both human and non-human, that resulted from the attack. Mark, that had to have left an enormous impression on you, and you've got to feel like you are part of the community at this point, I would think.

MARK SCHAMING: Some New Yorkers have told me that if I had lived here I wouldn't have agreed to go to Fresh Kills to be present during the cleansing. For the few of you who don't know, the New York State Museum in Albany is under the State Education Department. Its sister institution is the State Archives, the New York State Library. It's an old museum, begun in the early 1900s, that collects and researches the human and natural history of New York. So when September 11 happened, it was clearly something that the museum and the archives and the library were interested in documenting. In early October there was a very important meeting at the Museum of the City of New York. Some fifty or sixty people came together, probably thirty institutions and museums, and began talking about how to approach September 11. You know,

"Do we collect this? Do we take a photograph of and preserve some very basic things?" About a week later, a small group of us went into Ground Zero. We talked to the Port Authority about the ideas they had about collecting material. About ten days later, I, along with nine others, went to Fresh Kills for the first time for a tour of that site. We knew then that everything from Ground Zero was going there, that it was going to be a very important, unseen part of the whole story of September 11 and that it should be documented. So historian Craig Williams and I returned every week. Governor Pataki helped us gain access and eventually gave us tremendous state support to collect what we did. Over the next ten months, Craig and I went about twice a week and worked with the police and FBI at the landfill. They became our curators in the field. We basically collected all that we could, knowing that we would try to bring stories to these important objects. We also worked very closely with the FDNY. We had salvaged a truck from Managing Company Six, which is about a hundred yards from here around Beekman Street, and had worked very closely with the families, the FDNY, and that company to do that. In 2002 and 2003, we opened a major exhibition at the State Museum. We've had about three million visitors thus far through the exhibition in Albany. We probably have forty tons of material. There are five vehicles and hundreds and hundreds of objects that we collected from the towers and the aftermath. Things like the viewing platform that had been next to St. Paul's, scrolls of paper from Union Square, and on and on. We keep thinking, when the last thing comes in and the last person calls, that this is going to be the last object from the towers that we're going to get but then more arrive. In fact, this summer we took in an important collection from the FBI and a tremendous amount of memorial material from the FDNY. So it continues. A week doesn't go by where we don't talk to someone who has other pieces of the story to tell. We've been involved with the Tribute Center that's opening in a few weeks at Ground Zero. One of the great roles of a public institution is to provide service and to help share resources. So we're pleased that many of these important materials are slowly making their way back to New York, where they'll have great meaning.

JIM CAVANAUGH: I'd like get our discussion going by focusing in on a phrase that Julie used more than once when talking about the need for a sense of life. . . to preserve a sense of life and to center on it, because five years is a milestone. Whenever you have an event of great enormity, there comes a time when an event leaves the present and begins to become history. I don't know whether it's five years or not. I know Battery Park City. The authorities know we had a lot of people who were here on 9/11 and that some of them won't be coming to work this Monday, September 11, 2006, because they have too many memories and they just need to take that day off. For them it's still very much the present. About five or six months ago some military plane did a low fly over the area. Generally, when that happens, emails will be sent to the building, so that everyone knows there's going to be a plane flying low over Manhattan and it's okay. But that didn't seem to happen in this case, or at least the word didn't get out to this particular person in my organization. When the plane came by, it was low, and she saw it out the window, and she just burst into tears. It just brought everything back. For someone like that, five years is not five years, perhaps. As we talk about progress, a lot of it is couched in terms of what we built and we're going to build, which is very important, but we need to center as well on the human response. And so what I'd like to ask our panel is, do you have any thoughts as to where we are on that sense of life, and is 9/11 becoming history? Or is it still the present? Because moving forward in some shape or form is inevitable and important. Where are we?

TOM HEALY: I'll take a stab at that, I think it's a fascinating question. You know, 9/11 was the most photographed event in history. There are more people who witnessed what happened, particularly the second plane hitting and the buildings crashing and the reaction of the city later. There is an extraordinary archive of sound, of video, of still photography that comes pouring in from all over. If you think in terms of the other tragedies that America has faced, the amount of documentation of 9/11 is immense. What happened five years after Pearl Harbor, for example? There was gathering of about twelve Army officers with the flag, and that was it. There was no major commemoration, no major remembrance. Five years after D-Day? In fact, there was a complaint from the French president that basically said you've given us all these millions of dollars for the Marshall Plan and recovery but no one's come to mark this day with us. America fully lived the attacks of September 11 as much as the people in Lower Manhattan did. So the sense of trauma and the sense of presence that it has, I think, is unique in American history, particularly in the sense of identification. I think the possibility of it receding is different from other tragedies, because it's just too much part of all our consciousness. We will go through different stages in turning that into memory. We live in an age where time is compressed in different ways because of television. In a different age, building a memorial in the generation or two after an event would have seemed like an appropriate or even quick enough time. Now we look at five years and ask, has there been enough progress? Has there been enough change? And then the other issue that is very profound is that the attacks were deliberate, they were meant to happen in the middle of a thriving major city. This creates all different senses of need and time. It's not the fields of Normandy. It's not a remote port. New York is the capital of American commerce and culture, with millions of people living and working around it. So there are all sorts of pressures. Life has to carry on at the time we remember. It's a very complicated and emotional issue. I would just say that my own view is that there has been a lot of progress in five years. I'm not concerned that we haven't locked in a memorial—as if that could define what the 9/11 experience was. I don't think that that's been a bad thing. . . that we've had a longer process. It may not have been planned, but the result has given us more time to feel what that experience was.

CHARLES LAI: It's a very complicated question, because on the one hand I look at 9/11 on a personal level. I couldn't walk through the exhibit upstairs. I saw part of it, and I couldn't go through it. So I'm still living through that experience now. For the last few years, I have been finding myself going through the official ceremonial remembrances of 9/11. At the same time, for the last five years I've been doing 9/11-related work on a day-to-day basis. So, for me on a personal level, it's etched in my memory, and yet I try to move past it, not to say that I forget about it. Now, separating that from me as an individual, in my own community, in Chinatown, just as we're talking about in Lower Manhattan, and in America itself, we continue to live that experience. Life has changed for all of us. How do we adapt and take in the new circumstances? Chinatown, in particular, was devastated economically. I just want to give you some numbers. A few months after 9/11, a quarter of the adult working population was laid off. Many jobs were lost, lost forever. Many Chinatown residents work in the garment factories, in restaurants, and in service industries. Over forty percent of garment shops closed and closed permanently. We talked about how the garment industry was going to be on the decline. But like any industry, when it is going on the decline, there is a period of adaptation and change. Yet with this abrupt closure of doors, there was nothing for people to move on to. There was no training. More than seventy percent of the Chinese-American community in Chinatown do not have a high school diploma; nor do they have the English language ability. It is extremely difficult for any community living through that one-day experience, but having it linger in the following three or four years impairs your living; simply trying to figure out how to put food on the table or pay the rent becomes a challenge. So this is part of the lasting legacy in our community. Trying to find ways in which people can go on living. At the same time, the rest of America itself is trying to figure out how to continue. Everywhere else there's the question of how do we behave now? I don't quite know how we will be able to move it to the back of our collective minds. How can we say, "that was indeed a tragedy," then build a memorial, and move on? How can we say, it's over, it's done, our lives can be better, we're safe again? That's the big question, are we safe?

JIM CAVANAUGH: Julie, you mentioned you have three children. A lot of people in your position had decisions to make: do we want to continue to raise a family here, or should we leave, go to some place that might be safer. What were some of the factors that went into your decision to stay here?

JULIE MENIN: Well, I think Lower Manhattan is one of the best communities in the city, so for me there was never a doubt about staying here. I did not have children; I gave birth to my children after 9/11, so I *really* decided this was a community I wanted to raise children in, despite the fact that two years after 9/11, the EPA was doing air quality testing downtown and found forty times the recommended EPA level of lead in our apartment. So there are still ongoing risk factors for many people downtown, but I still come back to the fact that it is just a wonderful community, and particularly a wonderful community in which to raise children. It's interesting that the area where we live happens to be in one of the restricted zones. As you know, there are many streets downtown that are still closed to vehicular traffic. So in order to get to our particular apartment building we have to go through a canine checkpoint and a guard checkpoint if we drive. In the beginning, my three-year-old son would ask, "Why is there this dog here?" Now he's become so desensitized to it that he brings the dog treats. So I think things change. I think the legacy of 9/11 becomes a way of life for many people downtown. But I think that one of the reasons that Lower Manhattan is the fastest growing residential neighborhood in the city, one of the reasons why the population has doubled downtown in the last five years, is because it's surrounded by water. There are terrific waterfront amenities and recreational facilities. There's a real sense of a cohesive community, and that is why, despite some of the issues that occurred in the redevelopment, I feel very strongly people will continue to find Lower Manhattan a very attractive place to live.

JIM CAVANAUGH: Do you have friends who took the opposite point of view and left? And if so, how did you react to their feelings?

JULIE MENIN: Yes, of course. Many people I know did decide to leave, and I think that that's unfortunate. I think many of those people, I can't say for sure that all of them, but many of them, I think, are regretting it, because they see that life downtown really has rebounded. I don't want to say that there's a sense of normalcy, because I don't think there will ever be that. But it has certainly come back to some extent and I think there is some regret for people who moved out of this area.

TOM HEALY: They would certainly like to own their apartments.

JIM CAVANAUGH: You mentioned, Julie that you lived in one of the restricted zones, and that's something I wanted to bring up and get everyone's perspective on. It's something that we at the Battery Park City Authority feel somewhat strongly about. We have a mix of residential property and commercial property. The concern of the commercial property, whether it's the New York Mercantile Exchange or the World Financial Center, is that they continue to show up in target lists. So, they have a need for protection, and their response to that is that they want very visible protection. They like these rings of balers around buildings, and walls and barriers, because not only does it protect the buildings, but it sends a message that these buildings are protected, that if you're thinking of doing something bad, you should go some place else. Our point of view at the Authority is, we would like the look to be very different. We are trying to convince the corporate community that there are other ways to protect your building and still maintain the sense of a neighborhood rather than Fort Knox. One of the things we've done for example is work with a manufacturer to create this aerated concrete, which we're putting in a model installation now. What happens is if a truck goes into an area that it shouldn't be in, the aerated concrete will collapse and the truck will sink in. You put that around the building it means no one can get a truck bomb close to that building. It has the same effect as a ring of [balers] but it looks a whole lot nicer, because on top of the collapsible concrete you plant grass. The corporate community says well, it may work, but what if people don't know it's there? We want people to know that we're protected. So we find ourselves in the middle of this push and pull between what works for people who live here, and what gives a sense of an armed camp. Some kids have adapted to the whole thing very well. But I wonder as they grow up, is that a good thing? That they come to believe that this is just the way neighborhood should look. I'd be interested in everyone's perspective on what we've done to Lower Manhattan in that regard.

CHARLES LAI: In one of my multiple lives, I worked in city government. Around lunchtime, the workers around City Hall would walk through the courts then go into Chinatown to have a meal and go shopping on Canal Street in the jewelry district. Later, they would come for dinner. Now the workers won't venture past the armed guards or navigate the barriers. There's a sense of exactly what you said, Jim, that they're not welcome. Well, not really "not welcome" but that it's trespassing. That you're walking through your own city, but you feel like you are a trespasser and you could be potentially hauled away. So we are dividing our communities. One section of Lower Manhattan is literally off-limits. The life of communities in the city is that people can go and learn from each other. There are lots of things that Chinatown has to offer. The closure of the different streets, Nassau or Fulton or Park Row, is extremely unpleasant. People don't feel like they can walk through them.

JULIE MENIN: I think Charles has made an excellent point, particularly with the street closures around Park Row and the Financial District. This has served as a real impediment to cohesiveness in the community, and it's particularly hurt the small businesses. One of the things that our community board and community groups, as well as Chinatown community groups, have been very focused on is trying to urge the city to do more to mitigate the impact of street closures. The City Council did pass a bill on this, but I think the city needs to do more, particularly to protect the small businesses in Chinatown and Lower Manhattan who really can't survive these street closures. They could put up signs. There are cost-effective measures that could quickly be implemented that have not been implemented, and hopefully we will see that change in the near future.

TOM HEALY: I'd just like to tell a few stories from the point of view of some artists. Public policy issues, architectural issues, liability issues are something I can't speak to at all. But obviously many artists are engaged with this whole subject of what security is. Two years ago, there was intelligence, that turned out to be old, that the Stock Exchange and a few other targets in the city might actually be threatened. The particular date was near the September 11 anniversary. So of course there was all sorts of security, FBI, police with their submachine guns, etc. Not surprisingly, considering the American culture, all sorts of television crews showed up. You know, here's the bomb attack, let's be there to see someone drive a truck in. This is a very absurd notion. This was right at the Stock Exchange, around noontime. There were hundreds of militarized police and all sorts of tourists just milling around, to see what was happening. There were all sorts of television crews. We just happened to be doing a dance performance on the steps of Federal Hall right next to that. So it was, in a certain way, the quintessential New York moment of utter chaos. Well, the television crews had nothing to film—there was not going to be a terrorist attack—so we literally tapped them on the shoulder and said, "well look behind you, there is a dance performance going on." In an odd way, it became a critique of some of the security drama that was there, that people were literally doing something celebratory and creative in the craziness of that environment.

On the anniversary of September 11 last year, we did a show called "The Knock at the Door." It was a show in response to the Patriot Act. It turns out that many, many people have run afoul of new security regulations in the Patriot Act, particularly in immigrant communities. There is a great deal of concern about what may or may not be a danger. And we found out, just by scratching the surface a little bit, that in fact many artists, scores of artists, had been visited by the FBI, had their work confiscated, some had actually mistakenly been sent to jail and put on trial. They finally got off, but their lives were ruined. All sorts of things, because issues about security crossed the line into issues of freedom of expression and such. The show we put

83

on was a very compelling endeavor for us. We put all sorts of work up and left it up to the public to decide what were the supposedly dangerous ones. It was impossible to tell.

JIM CAVANAUGH: I want to open the discussion to the audience for questions in just one minute, but I want to answer one more question. We're trying constantly to bring the focus back to the human response to 9/11. Mark has seen it from a very unique perspective, talking to him I've learned how, originally, when they went to Fresh Kills, they were not greeted with open arms. Instead, their cameras were confiscated. Over time, the feeling of the authorities did change and they came to believe documentation of the attack was something that ought to happen. Eventually there was a change in attitude and a welcoming. Maybe you can tell us a little bit about whether you've seen a change in response to your efforts, or what the response has been, particularly from those who were personally connected to the attack.

MARK SCHAMING: That's a complicated question. There seemed to be a great need, right away, from the public, a kind of a hunger to know more about September 11. Objects are mute unless you know the history of them; they are the touch buttons of history. We have done as best we can to continue to talk with people. So you tell a story and use an object, that's what the museum does.

When we do exhibitions, and I know the World Trade Center Foundation is struggling grandly with this, we think about visitors. What are visitors going to take away from this? It's hard to pull away and say, "I think this is about that, and that's the story we should tell." It's interesting to me, when I see people from Ohio here at the site. What are they taking in? What is a New Yorker thinking about it? There are probably very different sorts of things that they're thinking about. So it's complex and difficult to tell these sort of stories. We've dealt with many families. They come to us, we don't call up people and say you've lost someone, do you have something? We've had a significant number of families step forward and give the museum some

84

things that they never want exhibited. A woman gave us her husband's wallet with seven hundred dollars in it that they had recovered from his body. She couldn't spend the money, she couldn't put it away, and she asked us, "could you take it," and we did. And we have it in the collection. We have other things that families want to have an individual remembered by. I think in time, when the museum develops here, the exhibit will probably change.

JIM CAVANAUGH: We've got about another fifteen minutes. I'd like to know if there's anyone in the audience who has any questions for any of us up here.

AUDIENCE: What is the vacancy rate downtown. Someone mentioned it was a double-digit vacancy rate down here?

JIM CAVANAUGH: I may have spoken about Battery Park's vacancy rate, after the attack on September 11. All of Battery Park had emptied out for a short while, and one building's people were gone for about a year. Our vacancy rate now is negligible. There was a higher commercial vacancy rate, but in terms of people wanting to live down here, our vacancy rate is very, very low. It's as low as anywhere else in New York. In some of the buildings, the developers tell us they actually have a waiting list of people who want to rent. Our very newest buildings are condominiums. We have one that won't open for another six months, and yet it's been sold out for several months. It sold out way in advance, really exceeding everyone's expectations. So our residential vacancy rate is very, very low. I don't know that anyone expected that five years ago. You might have thought no one would ever want to come down.

AUDIENCE: In terms of continuing to build life and culture in Lower Manhattan and trying to pick up the master plan, could each of you, from your own perspective, list the two or three most important things that you'd like to see happen over the next few years, both physically and financially, to continue and to build the community.

JULIE MENIN: Was your question with respect to the master plan, or putting the master plan aside?

AUDIENCE: I've just used the master plan in a generic sense. From your personal viewpoint, what do you think are the high priority areas that need to be pursued by the city and/or the community?

JULIE MENIN: Well I think one of the first priorities must be more schools downtown. As I said, our population has doubled in the last five years; it will double again in the next five years. We're in a real crisis situation with the overcrowding of our schools. If we do not find more school locations downtown, we're really going to be in a very difficult situation. That is, from the community board's perspective, our top priority. Another top priority is building the performing arts center, which of course relates to the master plan for Ground Zero. We firmly believe that this is a key component of the master plan, and if it could be built quickly it would actually serve to attract more office tenants. As to the gentleman's question earlier about vacancies, I was the one who mentioned double-digit office vacancies, not residential but office. There is some struggle downtown to attract more office tenants. If we had a beautiful performing arts center, it would serve as a magnet for the whole area; it would go a long way to attracting more office tenants. So schools and the performing arts center, as far as the community board goes, are really two top priorities.

CHARLES LAI: I would like to say that in the fairly near future, Lower Manhattan and the surrounding area will be the most visited place on Earth. But the infrastructure stresses that will accompany these huge visitations go beyond just the Ground Zero area. Chinatown is ten blocks away. The transportation system there needs to be upgraded. We want to have connections between communities so that Ground Zero is not an isolated location, so Battery Park City is not an isolated location. Lower Manhattan is a very dynamic area, there are many,

many museums in this entire area. How do we make certain that there is enough of a flow for people to venture from one part of the city to the other? We say this is the cultural capital of the world; well, only if you can access it, only if you can partake in it. The richness that each one of these communities and organizations and museums can offer is extraordinary. The thousands of artists in this area. . . there's so much to offer. The master plan must address ways in which people can get around, get around nicely, and also be able to connect to the local businesses. After a nice performance, people should feel they can go and eat, go and shop. We're not only talking about mom and pop stores, but all the small businesses in this area. They have to be supported. How do we make those things happen?

TOM HEALY: There are three things that I would like to see. First and foremost, though, I would like to echo Charlie's statement that small businesses, restaurants, and stores are the lifeblood of community in Lower Manhattan. There is not in the master plan a significant effort to support those small businesses, and I'd like to see something that addresses them across the board, whatever those kinds of businesses are. The second thing, is because of all the construction, the street closings, the security and such, what's not part of the master plan is an effort to push through projects that ameliorate those conditions. There are going to be all sorts of barricades and shrouds on buildings, trucks, and noise. I'd like to see a range of projects that make that more livable through art. The third thing that I would like to see, and this is a little more parochial to the small arts organizations, but I would like to see an extensive marketing effort. Below Canal Street there are over a hundred small cultural institutions and non-profit institutes, and I would like to see a public-private partnership in a significant marketing effort to draw people back during this period of construction and confusion. I think there's an expectation that people will wait until after the master plan is complete, but I think we need to concern ourselves with a lot of the transitional period.

CHARLES LAI: If we wait, it will be too late.

MARK SCHAMING: I guess it's no surprise that I would think that the most important thing is the establishment of the museum. I think it's going to be, if possible, one of the great American museums at the most powerful site in American history. It just seems to be, and will continue to be, dogged by complicated political pressures. I would just hope they let the museum experts do what they know how to do. I think it has to be distinguished from the memorial area. Michael has done a great job with what he's doing, and I know that's complicated too. I always want to hear of both the memorial and museum when they're talking about the master plan for Ground Zero because they're both very important.

TOM HEALY: If I may, just to echo Mark, I don't think it's gotten a lot of press in the controversy about the memorial, but there's been a remarkable team put together to build this museum. There already are some extraordinary curators and a director, and I think it will be an extraordinary museum when we get it built. They are a team that will make it one of the great museums.

AUDIENCE: I live in Battery Park City, and before 9/11 one of the great things about living down there was the fireworks along the Hudson River. After 9/11, they're great if you can see them, but they're frightening if you can't. I mean for those of us who live there, if you can't see the lights, it can be incredibly terrifying. If we can't get the Coast Guard to understand this, how are we going to get somebody from Ohio to understand?

JULIE MENIN: That's an excellent question. It was the community board who first contacted the Coast Guard. We've had numerous meetings about this and they promised us that there would be proper notification. There have been a few instances this year where there has not been proper notification. In cases like this, please call the community board and let us know, and we will complain, because we've set up a system where there is supposed to be proper notification. If that is not happening, we would want to know about that right away.

AUDIENCE: I'm a thirty-five year resident of this area. This had long been an isolated area. Before the Seaport was built, I used to have to tell a cab driver that I was pregnant and I needed to go to the hospital in order to get a driver to come down here. Over the course of the years we've seen some remarkable changes. I realize that there's a need for security, and I'm so happy to hear about the brilliant possibilities of having some of those security issues addressed in a more pleasant way. But what concerns me is that there's always been a middle income group here and a low income group, and it seems to me that the economic impact of the focus on this area will drive those people away. I'm wondering if anything is being done to address a mixed-use community. To address the needs of the people who are not earning tremendous salaries. I'm about to retire, and that concerns me. I think that more people come down here because it's less expensive, I think the pocket-book is what's driving people.

JULIE MENIN: I'm happy to address that. As you probably know, the Lower Manhattan Development Corporation had allocated fifteen million dollars to address the issue of affordable houses, but that money, unfortunately, has not been spent, though we've been very vigilant about trying to get them to spend it. I agree with you, there's a massive influx of luxury housing downtown, but not enough housing to support middle income and lower income people. We're very attuned to this issue, particularly in re-zonings. In North Tribeca, right now, we're working on a re-zoning and we're asking City Planning to put in an affordable housing component. I completely agree with you and it's something that the mayor has said that he's very committed to. We need to build more affordable housing downtown so we do not price our school teachers, our firefighters, our police officers out of this community. Because that really would be a tragedy.

AUDIENCE: I'm a local resident as well and five years later as I reflect back on 9/11, the schools seem to be the issue for those of us raising families here. Besides that, recreation is a serious concern, as well. But what I haven't heard mentioned today is the health concerns that we're dealing with. It's a very serious and pressing issue, and I do have children that are suffering from some of the respiratory concerns.

JULIE MENIN: I'm happy to answer that question. In fact, we had a very big victory yesterday for the community. The community board asked Larry Silverstein to retrofit every single construction vehicle with ultra-low sulfur diesel fuel and he agreed. This is going to result in ninety percent decreased emissions which is so important since diesel fuel has been conclusively linked to cancer and asthma. We all know that the health of our children and residents downtown has already been compromised, post-9/11; they're very susceptible to respiratory illnesses. The issue of health is so important; we really hope that a clinic is set up soon that will *treat* individuals. Right now the WTC Health Registry is monitoring residents' health but they're not going to be treating the residents. If it is found that there are cancer clusters or other serious health effects downtown, we need to set up a mechanism now to treat them, and certainly to pay for those who are not insured.

JIM CAVANAUGH: Battery Park City has already begun addressing issues of health and the environment. We put forth sustainable building guidelines prior to 2001 and we have green buildings. Part of the green building guideline is that you must provide very good indoor air quality. Again, these requirements were put in place by the Authority before 9/11, because we felt that it was something that needed to be done in response to the general quality of air in urban areas. Since then, it's become very important. We find one of the things people really do respond to is the fact that in our newer buildings, all the air is filtered. There are two filters, one in the public spaces in the building and then a secondary filter in each apartment. So while you can open the windows if you want, a lot of people never do.

It's become a very, very strong selling point for people who want to live in this area.

We have time for one more question and then I'm going to ask the panelists if they want to make any concluding remarks.

AUDIENCE: Now everyone is sitting around planning things and it all looks very nice, but when are you building this art center, to reach out to art, New York art, art that was born and lived here for how many years?

JULIE MENIN: I'm happy to address that. I think we have on our panel an excellent example—culturally—to honor artists in the community exhibiting their work. They have an artist-in-residence program. It's not just about bringing in new institutions, it's about really supporting the artists in the community and helping them.

JIM CAVANAUGH: We recognize that we have an obligation to the arts, which is why we have a number of public art installations throughout the neighborhood. We give a very significant amount of money from our annual budget to various arts organizations and we put on a lot of performances; I mean if you think about how many performances there are here, not just in the course of the summer, but year-round, it's a lot.

And it is a range of artists, it's not always the particularly well-known ones. I think there's an effort to reach out and help fund artists that are not particularly well-known or those who are just coming into their own. That effort does go on, but we can always do better and we certainly appreciate your perspective, because despite all this organization does, I'm sure they'd always like to reach out even more. I'd like at this moment, to ask if anyone has any final wrap-up comments.

TOM HEALY: On the air quality issue, we've actually commissioned a group called Preemptive Media to look at this whole issue. They're creating what they're calling a social sculpture, which is engaging thousands of people who live and work downtown and creating individual air testing kits for people to check their own air quality, where they live, with their families, with their children. We're compiling a big database of what people find so that citizens have a sense of control of that information and participation in it.

In terms of our commitment to the whole range of diversity and to the artists who work here and the people who aren't necessarily going to go to the ivory tower expensive performing art center, we continue to look for ways to be better about that. We just did a project called Harlem Is Downtown, bringing everything from gospel choirs to hip hop groups to Lower Manhattan because the African in America began at South Street Seaport.

CHARLES LAI: Community is many, many things. It's not just America, or simply the community below Canal Street, or Houston or 14th Street. I think there are a lot of things that we need to figure out. It's important to look at what we want our future to be, a future that has the truly diverse communities to which we cling, from the more affluent, to the business people, to the artists. All of those people have community, they all have their special requirements and special needs and the question is how do we insure that strengthening one sector of the community is not done at the expense of the other. In terms of working class communities in particular—at least in Chinatown—we must ensure that they're very much part of the overall planning. And I know planning doesn't mean anything without the money to be put into that effort. So I want to say that we have to put the money there.

JULIE MENIN: Okay, I'm going to be very brief. Hopefully the factions and politics that have caused problems with the rebuilding will be put behind us and in the next five to ten years we'll really see the site rebuilt and the surrounding areas rebuilt, especially in terms of building more schools and affordable housing and protecting our small businesses. Charles's point, very aptly talked about, is bringing all the communities together so that there are no barriers between, for example, even crossing the West Side Highway to go to the financial district, or going from Tribeca to Chinatown, is essential. Those barriers have physically divided the communities and hopefully that will change and we'll really create one cohesive Lower Manhattan community.

JIM CAVANAUGH: I'll make it really short. When moderating a panel, one of the hard things to know is how difficult it's going to be to keep it going. Thanks to the interest and following of everyone here today, it turned out to be, I think, an incredibly easy job. Also, thanks to people in the audience, we had some great questions, you really made this whole thing work. I want to thank Pace for having sponsored it. There's still a lot left in terms of the two day conference, but I think they should be congratulated for pulling this together. I think it's turned out to be beyond anyone's expectations and I want to thank you all for coming. Thank you.

KEYNOTE SPEAKERS

Left: *U.S. News & World Report* Editor-at-Large David Gergen visits Pace Professor Darren Hayes's Technology Systems class.

Right: Daniel C. Doctoroff, New York City Deputy Mayor for Economic Development and Rebuilding

Left: Vice Chair of the 9/11 Commission and member of the President's Homeland Security Advisory Council, Congressman Lee Hamilton speaks to students in Pace Professor Chris Malone's Political Science class.

KEYNOTE SPEAKERS

Right: Presidential
Historian Doris Kearns
Goodwin and Pace
University President
David A. Caputo

Left and Below: William
Kristol, Editor of *The Weekly
Standard* and co-author of
*The War Over Iraq:
America's Mission and
Saddam's Tyranny*, speaks
to the Pace University
Young Republicans

PANEL ONE PARTICIPANTS

Left: The Michael Schimmel Center for the Arts at Pace University, site of much of the Aftershock conference.

Right: Jim Dwyer, *New York Times* reporter and author of *102 Minutes: The Untold Story of the Fight to Survive Inside the Twin Towers*, and Michael Emmerman, Director of The Special Operations Support Group

Left: Panel Moderator Joe Ryan, Professor of Criminal Justice, Pace University

PANEL TWO PARTICIPANTS

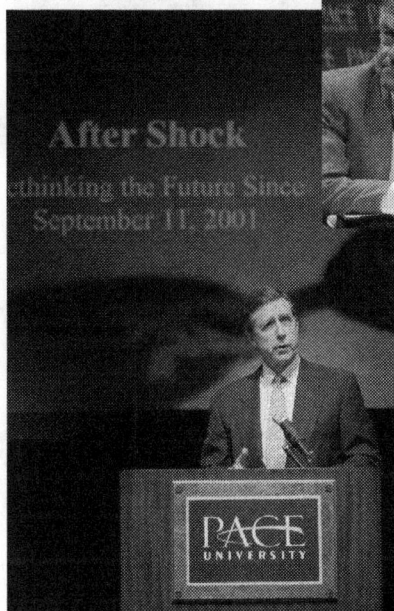

Above: Joseph Baczko, Dean of the Lubin School of Business, Pace University; Joe Petro, Citigroup Security and Investigative Services; and Michael Dolfman, Bureau of Labor Statistics

Left: Joe Petro, Citigroup

Right: Eric Deutsch, President, Alliance for Downtown New York

97

PANEL THREE PARTICIPANTS

Right: John Cronin,
Director of the Pace
University Academy for
the Environment,
standing in front of a
photo of 9/11

Left: New York
Congressman Jerrold
Nadler

Right: John Cronin,
Lorna Thorpe, David
Newman, Bruce Logan,
and Congressman
Jerrold Nadler.

PANEL FOUR PARTICIPANTS

Above: Julie Menin, Chairperson, Community Board 1, and Jim Cavanaugh, Panel Moderator and President and CEO of the Battery Park City Authority

Right: Mark Schaming, Director of Exhibitions and Programs, Curator, New York State Museum, "The First 24 Hours" Exhibition

Left: Tom Healy, President, Lower Manhattan Cultural Council, and Charles Lai, Executive Director, Museum of Chinese in the Americas

PANEL FIVE PARTICIPANTS

Panelists (l to r) Tom Roger, Sally
Regenhard, Gene Steuerle, and
Richard Shadick, panel moderator

PANEL SIX PARTICIPANTS

Left: William Rudin, Chairman, Association for a Better New York, and Robert Yaro, President, Regional Plan Association

Right: Kathryn Wylde, President, The Partnership for New York City

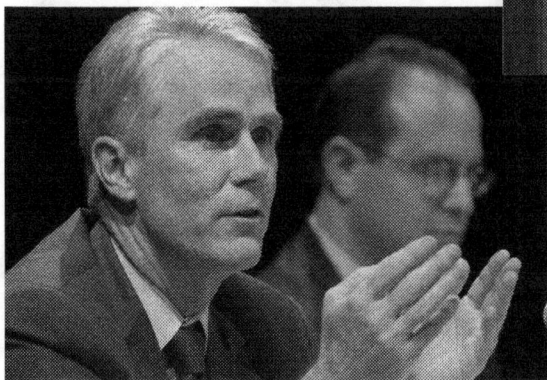

Left: John Cahill, Secretary to Governor George Pataki, and Stefan Pryor, former President, Lower Manhattan Development Corporation

PANEL SEVEN PARTICIPANTS

Left: John Merrow, Producer, NewsHour with Jim Lehrer; Robert Hackman, Graduate Student, Pace University; and David Warren, President, National Association of Independent Colleges and Universities

Right: Robert Hackman, Graduate Student, Pace University

Left: Sadie Bragg, Provost, Borough of Manhattan Community College, and David Caputo, President, Pace University

PANEL EIGHT PARTICIPANTS

Right: Steve Clemons,
Director, American
Strategy Program, New
America Foundation

Left: Beverly Kahn,
Panel Moderator and
Vice President of
Academic Affairs, Pace
University

Right: Panelists (l to r)
Alice Greenwald, Beverly
Kahn, Ken Jackson, and
Nikki Stern

EXHIBIT: THE FIRST 24 HOURS

Left: Mark Schaming (r), Curator of the NYS Museum exhibition "The First 24 Hours," photographed during the exhibit's installation

Below: Member of the FDNY viewing a timeline of 9/11 which was part of the exhibit, "The First 24 Hours"

ARTIFACTS

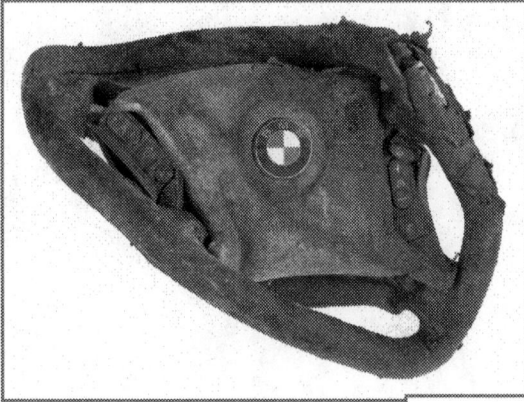

Artifacts recovered from Ground Zero and the Fresh Kills station, on display as part of the NYS Museum Exhibit, "The First 24 Hours"

PANEL FIVE

THE VICTIMS' FAMILIES AND THEIR INFLUENCE ON PUBLIC POLICY

RICHARD SHADICK: Good morning. My name is Dr. Richard Shadick and I'm the director of the counseling center here at Pace University. I'm also the director of the trauma response service of the William Alanson White Institute. I have the pleasure of being the moderator for today's panel.

It has been five years since the terrible disaster that shook our world and destroyed the very foundation upon which we stood. Perhaps we understood then that the events of 9/11 would be influential. Certainly, though, we did not fathom its full impact. In psychology, for example, the disaster lead to the reworking of theories and a new understanding of loss, grief, post-traumatic stress, and the resilience and growth that it fosters.

While the title of today's panel is "Victims' Families and Their Influence on Public Policy," this session is about survivors and resilience. I witnessed resilience moments after the towers came down as the Lower Manhattan workforce escaped the gray dust debris by seeking refuge in the halls of Pace University, people of all backgrounds and means picking up, dusting themselves off, literally, and moving forward. I saw it unfold in the subsequent months with Pace students and colleagues when I tended to those who witnessed horrible scenes on that morning's commute. I saw it in the Ground Zero rescue workers I counseled who had worked tirelessly, compromising their own health to clear the ruins. And I saw it in the New York City public school teachers I consulted with who had to deal daily with the scores of traumatized and confused children.

I saw resilience then and I see it today in my clinical work with individual survivors who lost loved ones either from the towers' collapse or from the emotional sequelae of the day's events. Their resilience, their will to move forward has led to growth. However, that growth is not universal and there's still much healing that needs to be done.

Witness this morning's newspapers, particularly the front page of the *New York Times*. A *New York Times*/CBS poll found that 72% of New Yorkers and 50% of the nation feel that the government needs to do more to protect us. Two thirds of New Yorkers are very concerned about another attack, a slightly lower level since 9/11. These responses and others from this interview demonstrate the need for more healing.

Today we will hear from resourceful and resilient individuals who have not only recovered but have become stronger and have had a positive and lasting influence on public policy. After their introductions, I will pose a series of questions to them and at the end the audience will have a chance to ask questions. However, before we begin I would like to pose a question to the audience, something they can think about as they listen: What can we, as concerned individuals, do to promote the resilience and growth that seems to have eluded us on this anniversary?

I'd like to start by introducing the panel members. We have the founder and the chairperson of the Skyscraper Safety Campaign, Sally Regenhard. Next is Tom Roger, the co-founder of Families of September 11 Incorporated and the director of the World Trade Center Memorial Foundation. And finally we have the cofounder of Our Voices Together, Eugene Steuerle.

We'll start with Sally.

SALLY REGENHARD: When I and my family founded the Skyscraper Safety Campaign, shortly after 9/11, it was not our intention ever to influence public policy. The purpose of starting our organization, first of all, was to express our outrage that something like this could happen to this country, to this city, to my son, and to the nearly three thousand people who met a brutal and needless death. As time went by, the purpose of educating the public about the true issues of 9/11 really became the main factor.

My son was a probationary firefighter who graduated from the fire academy only six weeks before 9/11. He was one of seventeen probationary firefighters and one of ninety-seven unmarried firefighters who were lost at the World Trade Center. His energy, his beautiful life, inspired us to look further to question how this could have happened. Why were New York City, New York state, and this country so unprepared for another terrorist attack, despite the fact that the Port Authority was the victim of a terrorist attack in 1993? As time went by, we found out about the failures of this city, state, and the Port Authority in emergency preparedness, emergency planning, and emergency management.

So many people died that day because of the failures of communication both in the firefighter radios and the collapse of the 911 system. These are the issues that we sought to bring to the public attention. We went to Washington and we asked for a federal investigation into why the World Trade Center collapsed and all of the other issues. We did achieve that. We have sought to educate the public about the need for increased building codes and fire codes and certainly we continue to advocate for the public to have all the information about how this happened to us.

Actually, something that I'm very proud of is that my organization along with my wonderful civil rights attorney, Norman Siegel, succeeded in winning the right to disclose all of the 9/11 tapes, transmissions, and emergency calls. We fought for three years in court for this. When we went to the New York State Court of Appeals it was an honor and a privilege. We're very, very loyal and we're very patriotic people. The truth is the most patriotic thing that you can have in your life.

RICHARD SHADICK: Thank you, Sally. Tom?

TOM ROGER: Five years ago, I was one of those people who watched from afar from my office in New Haven as the horror occurred on the morning of 9/11, not realizing that my twenty-four year old daughter, Jean, had agreed to stand in for a co-worker on American Flight 11, about ten minutes before it took off. And so, as I was thinking what a terrible thing was happening to those people in New York I didn't realize until the second tower had collapsed and my wife had called me that in fact I was one of those people in the dark cloak that descended on all of those families. I became one of those families.

This type of sudden and emotional loss can strike at the very foundation of your faith. Even our minister said to us that what happened on 9/11 made him question his very faith. The ground that you thought was somewhat stable suddenly becomes very unstable.

So, in the aftermath of what I would call a crushing personal tragedy even while the world was mourning our losses, some of us became motivated—Sally just mentioned her story and Gene will tell you his—to influence public policy. My story is similar to Sally's story in that I was outraged because I had some knowledge of the issues that led up to a number of people dying who shouldn't have died. The first issue in which I had some background was airline security. Just days after 9/11 it seemed to me that somebody, amidst all of the confusion that was going on—you know, whose fault was it and why had this happened, and lots of public officials kind of running for the woods saying that it wasn't their fault—somebody at least ought to be trying to shine a light on what some of the issues were, try to raise what seemed to be some fairly common sense questions about things that ought to be done.

So, with the help of some people that I actually happened to be working with at the time from Hill and Knowlton International, which is a large PR firm, along with some other airline experts, I authored an article that got published as the viewpoint article in *Aviation Week and Space Technology* on October 1, 2001. In it, I called for specific recommendations in eight areas. And again this was not me thinking, as Sally had mentioned, that I had any ability to influence public policy. It was me trying to get something off my chest and expressing some anger and outrage at what was happening.

The interesting thing was that during that timeframe, if you'll recall, there was a tremendous amount of pressure on officials to try to do something. And the problem was that all of the normal sources of what should be done had become somewhat contaminated by what happened. Leading up to 9/11 there had been low level discussions about airline security but nothing had been done. Of course, the airlines didn't really want anything to be done because it might affect business or cost money and the FAA was pretty much in their camp. What I was able to do, what some of us have been able to do, is stand up and shine a light on issues that need attention and make things to happen. Sometimes it's called moral authority. I hesitate to use that because I'm not sure we necessarily have any particular authority, but we do have a voice. As with the Mothers Against Drunk Driving, there does come a time when the families who are most affected should have a say in what happens with public policy.

The aftermath of that experience with the *Aviation Week and Space Technology* article was that the legislation that was passed on November 19 actually incorporated most of the recommendations that were in my article. And it's funny, because when I followed up with some of the congressional people after the legislation was passed, I said, "That legislation looks vaguely familiar." They said, "Yes we liked what you said in your article and in fact it came across as being so unbiased that we had to follow that particular viewpoint."

So again, I never got into it from the standpoint of wanting any sort of involvement with public policy. It simply became obvious to me, along with some of the other family members, that if we stood up and tried to take a no-nonsense approach to some of these issues people would listen. I was fortunate to meet up with a group of other people who felt the same way. We founded The Families of September 11. The founding board of Families of September 11 consisted of six lawyers and two PR people and one lobbying professional. These were people that were not afraid to stand up and take on public policy issues in a very serious way.

We've had some great success. We took issues with the victim's compensation system, airline safety, security and emergency preparedness, and many other post - 9/11 issues, such as the teaching [about] 9/11, taking care of the children of 9/11, and supporting efforts such as Sally's Skyscraper Safety Campaign. We think it's a wonderful program that she's had. Many of the family groups joined together with the independent commission to investigate what happened on 9/11. It was a very successful

111

effort and we've been trying to follow up on all the recommendations in that area.

This has not been my full time job. I do it in memory of my daughter, who had a great life, and in memory of all those who were lost. I feel it is our obligation to try to make something positive happen.

RICHARD SHADICK: Thank you, Tom. And now our panel opener with Gene.

GENE STEUERLE: I'm the chair of an organization called Our Voices Together, which is a network of families, friends of families, and individuals like yourself who have tried to bring compassionate outreach to needy people around the world. We do this not only to try to meet their needs but as a way of sending a message different from the terrorists.

What unites you in the audience with those of us on the panel is that five years ago we were all given a voice and a perspective that we didn't have before. And for those of us on the panel the voice we've gotten is not one we really wanted, obviously, certainly not one we think we even necessarily deserve. But we concluded, one way or the other, that this new voice was something we had, it was an asset, and we had to decide how we could use it most effectively. But in most ways the goals of us who are directly involved with the families or were friends of the victims aren't really very different from yours. We're trying to find some way to bring some good out of the evil of 9/11. For those of you who are religious, you might say we're trying to do something redemptive.

The very first action taken along these lines was by you when you reached out in an extraordinary compassionate manner to those of us who lost family in the attacks of 9/11. None of us think of ourselves as special because of having suffered a death. There's probably no one in this audience who hasn't had a death in the family that grieved them. What we did receive at that time was an extraordinary, extraordinary outpouring of support which reflected, I think, your desire to reach out compassionately to us. We've also concluded that the effort that you made at that time was not just simply because you wanted to be compassionate. It was because you wanted to be empowered to do something to act at that point and time, and to act in the future, to really do something to fight against terrorism. And you did do it. You not only gave help to us, but you actually came forth in a way that empowered yourselves. You sent a message about what people are like and that message, and I know I speak for all the panelists, is one that gave us extraordinary hope at a time when it would have been very easy to despair.

I want to now turn to Our Voices Together which is a group of families that's trying to take that compassionate response you gave to us and reach out in a compassionate response to others around the world. That is this voice that you've started. It is a voice we think can rise up in the response to terrorists. It's the voice of ordinary people who realize that they don't have to stand on the sidelines. That you and I really can make a difference. And not only can we make a difference in the lives of others but we can send forth a very different message from that of the terrorists who send one of fear and despair. I would love to tell you everything I know about this wonderful and growing group of people, but in the interest of time I'm going to give a very shorthand version and simply tell you about some of our members.

• We have people involved such as Susan Reddick and Sally Quigley, two 9/11 widows who do bike rides to raise money for widows in Afghanistan.

• Liz and Steve Alderman, who lost their son, dedicated the resources they got after 9/11 to building trauma centers in places like Cambodia and Uganda and other places affected by terrorism.

• Leslie Whittington was a professor at Georgetown who, along with her two children and husband, was killed on the same flight as my wife. Their friends, David Stapleton and Joyce Manchester, decided that as a couple they needed to reach out. They've given to a whole variety of causes, including midwives in Afghanistan.

• Don Goodrich, a colleague of Tom's on Families of September 11, and his wife Sally have responded to the death of

113

their son by building a school in Afghanistan and want to build more schools. They want your help.

• After their brother and son, Jeff, was killed, the Gardner family set up a scholarship fund for college students to build houses in Latin America and other places.

What we've done in Our Voices Together is tried to gather together these families and these friends of families to promote the activities they have engaged in and also offer opportunities for you and anyone else who wants to engage in similar activities. So we're involved with Global Giving and Save the Children and a variety of other organizations. Anyone who really wants to respond in this way can really send a different message. We can jump to the film now.

[AT THIS POINT, GENE STEURLE RAN A SHORT FILM CLIP. BELOW ARE SOME OUTTAKES FROM THE SOUNDTRACK]

"If you're given something, whatever it is, a trial, a tribulation, money, friendship, use it. You know what responsibility does. Bring what good you can out of it."

"We are all now joined together by a common idea, a common vision, a common goal, and this gives a great lesson, a great message, to ordinary people elsewhere, that you can do something. It empowers people. You can change the world. And we really need to change the world considering what brought us together in the first place."

"My mom was a very energetic very forceful woman she was also incredibly kind and I think that she used her energy and focus in such a way that she tried to help people. After my mom died and we learned that we were going to receive money from the Victims Compensation Fund we sat down as a family and talked about what we wanted to do with the money."

"We realized, just because of the outpouring of sympathy and everything else, that we really did have a voice. And so then the question arose what can we do with that voice and how can we use that in some sort of positive way. And we thought that if we

took this voice, combined it with those of other families and friends and really just other concerned people, that it might create a ground swell to do a lot more good."

VOICEOVER: Gene Steuerle and his daughters decided to start Our Voices Together a non-profit organization dedicated to supporting and encouraging international development projects and fostering goodwill and understanding through education and dialogue. Following are testaments to the contributions of Our Voices Together.

"It was very in line with something that Mom might want to do. It was inspirational, it made me feel better about losing her to try to make her proud and live up to her example."

"We don't want my mom to be honored because of how she died, we want my mom to be honored because of how she lived."

"Our Voices Together is first trying to reach out to people around the world saying look we care about your education, we care whether you're fed, and we're willing, as families and friends of victims, to devote substantial resources to this. And our goal is not simply to help you, our goal as an organization is also to create goodwill throughout the world because in the end I think it's goodwill among people that can create peace."

"Across the countries, the Steuerles found families and friends who had lost loved ones to acts of terrorism. And who shared their vision."

VOICEOVER: A friend of Peter Goodrich's who was stationed in Afghanistan wrote to Sally Goodrich and explained about the desperate need for education and supplies. With the help of their local community, Sally and her husband Don have raised nearly $200,000 to build a school for girls in Afghanistan. In the spring of 2005, Sally traveled to Afghanistan to see firsthand the progress being made on the building of the school.

115

"Peter was intellectually curious, he was funny, he loved life, he was generous to everyone, and my job is simply to follow his lead and I haven't felt so close to him since 9/11, really until we began this project."

"Even creating one small education institution, just two small rooms, begins to change lives. Hundreds of lives."

"I was the recipient of great warmth and the hospitality is really hard to describe. From the moment that you step off of that plane they feel responsible for you and in every way they care for you."

"That portion of my life which was a void, a never-to-be-filled void, has now become much more of an urge to act in a way that he would have. And I know what it is that my child would have done himself and I know exactly what he would want me to do. He would love what I was doing and he would be beside me all the way."

VOICEOVER: In 2002, David Stapleton and his wife Joyce Manchester felt compelled to make a larger donation than they ever had before to support the Safe Motherhood Initiative in Afghanistan in honor of their friend and colleague, Leslie Whittington.

"I wanted to do something that I thought would be meaningful to Leslie if she could see it. And part of her professional life was devoted to looking at the relationship between work and family for women in this country but also in developing countries."

"I think if Leslie were here, and Charlie for that matter, that this is the sort of thing they'd be doing too. It just seems to me that this is the way Americans can respond to terrorism. It's a very constructive way and also a way that will reduce the fact of terrorist acts in the future. You know, I realized after hearing about the program that the relatively small amount of money that we were contributing was probably going to save hundreds of lives just over a very short period. It's really incredible."

"By choosing particular projects we can make a difference to a few people in the rest of the world and if those efforts are multiplied enough times we can really make a difference throughout the world. And it's a terrific way to honor people who lost their lives to acts of terrorism; it's a way to carry on the work that they would have done had they lived."

VOICEOVER: Susan Reddick and Patty Quigley were both pregnant when their husbands were killed on September 11, 2001.

"I was married to my college sweetheart, David. We had two children and I was pregnant with our third. David left home very early on that morning to catch a flight to Los Angeles. He left notes on the breakfast table for the kids saying that he would miss them and to have a good day at school."

"From the minute I found out that Dave had been killed on September 11, we were inundated with help and support."

"Because of all of the support from the nation and from the neighborhood and from just the community I felt like I needed to give back and to help. I thought, 'Wow. Maybe I could help one woman in Afghanistan and give her financial support and even maybe communicate and give each other emotional support.'"

"Patty turns to me and says, 'We should ride our bikes from Ground Zero back to Boston.' I looked at her and I said, 'Great idea Patty!' And so then we ran out and bought bicycles."

VOICEOVER: On September 9, 2004 Susan and Patty embarked on a 3-day 275 mile bike ride from Ground Zero to Boston raising $150,000 for widows and children in Afghanistan.

"Once I dried my tears and got on that bike it was an incredible feeling to pedal away and literally move beyond 9/11. The people's lives in Afghanistan have directly affected our lives here in the United States and the disparity between what we were

117

receiving here and what the women in Afghanistan were not receiving was just so huge that we just felt this connection that we're widows here, they're widows there. We love our children, they love their children. We want our children educated, so do they. We really all want the same things."

VOICEOVER: Susan and Patty hope to raise even more money for widows in Afghanistan. And they will be joined by hundreds of cyclists throughout the world.

Our Voices Together address both national and international needs with the aim of sending a message of compassion and global community.

"Everybody has something special that they want to help with somewhere in their lives if you could just put a small little part of your heart aside to the women in Afghanistan and let Susan and I draw you in and help us support the women, it will go a really long way."

"This is a passion of ours, that we feel we could make a difference, that we think it would make a difference in the world."

"With our collective voices we hope to make a little better place in the world."

"I care about the outcome being right for the world. I care about taking some kind of action that will create a different environment than the one that existed on 9/11."

"Not only are we remembering the lives of those who died but we're also responding in a way that makes a difference in the way that they would have wanted to make a difference."

"Overall, I hope that we'll find a way to make a little bit more peace in the world somewhere."

"Our idea was to use our voices, because of who we are, to get people to take actions that would make this a safer world."

"I think what Our Voices Together is doing is just amazing, that it gives a place for people to go when they say, 'I want to help but I don't know what it is I want to do.'"

"Transcend the personal tragedy into creating a vision of a world in which we can reach out to others, a world which is

inclusive, a world which is understanding of dialogue, of accept-
ance, and ultimately of compassion."

"You don't choose the role in life that you're given but you can
choose how to play the role that you're given."

"And I believe that the kinds of people that are giving us this
kind of response in the 21st century are going to be the martyrs
for other people to be looking at and to follow. That's why I find
that what Our Voices Together is doing so inspiring."

* * *

GENE STEUERLE: Please think about joining us.

RICHARD SHADICK: Thank you very much. That's a wonderful
example of the resilience and growth that we were talking about
a little bit earlier. What we'd like to do now is have a couple of
questions and answers from the panelists and a little later on
open it up to the audience.

I know there is one question on my list that all the panelists
had talked about wanting to have a chance to get a crack at, so
let's just go to the first question which is, What is the single
most important change needed in governmental policy today?

Want to start, Sally?

SALLY REGENHARD: In my opinion, the single most important
change that's needed in governmental policy today is to achieve
accountability and responsibility from all levels of government,
from the federal to the state to the city of New York, including
the Port Authority. The response to 9/11 has been characterized
as a failure of imagination. I don't agree with that. It was a failure of people to do their jobs on every level of government from
Washington to New York City. If we don't insist on accountability and responsibility, there will be no impetus to change any of
the policies that we have.

The only person who ever apologized to the families of the
victims was Mr. Richard Clarke, who testified in congress and
turned around and apologized to the families. We need to have

119

more people be accountable and responsible. The public has to insist on it.

TOM ROGER: I agree with Sally on the domestic front. On the international front, possibly for us to avoid continuing to serve as a target for every terrorist organization in the world, maybe we could have some new sense of diplomacy that doesn't include starting with guns first. That would be my wish for what could happen.

GENE STEUERLE: Well, while not trying to assess blame, I'm sure that on some level there is a lot of blame that could go around. I think after the end of the Cold War the United States in many ways retreated from its role in the world. Our assistance abroad is one percent of our national income and less than one-tenth the level it was when we engaged in the Marshall Plan. We were a much poorer nation at the time. Public diplomacy has been abandoned and I think the whole strengthening of the international apparatus of our government from public diplomacy to foreign aid to making that aid effective and getting it to individuals and not just to governments is probably among the most important reforms of government policy that's required.
Thank you.

RICHARD SHADICK: What personal factor or experience was instrumental in allowing you to overcome your own loss to powerfully influence public policy?

GENE STEUERLE: Well, I'll give a short answer because I actually said a little before. I think what helped us more than anything was you, you people in the audience and you people who are listening, who responded to us, and just gave us, right from the start, an extraordinary amount of hope. As I said before, anybody who has suffered death knows that there are moments of despair. And you really gave us hope that the individuals were really good and good-willed and that they wanted to do things. That, I'd say, more than anything else, is what really gave us strength from the start. Your feeling empowered to respond to us gave us sort of the energy to try and be empowered to do something ourselves is what most influenced me.

TOM ROGER: I mentioned earlier that the group of us who managed to come together to form Families of September 11, along with a number of other family groups, including some of the people sitting up here with me today—their sense that the world can be a better place, that it doesn't have to look like what happened on 9/11 for the rest of our lives and children's lives is empowering. You can work hard at some of these terrible challenges and you can make good things happen. So I think the sense that there is some reward that could come from becoming activists in these areas was what allowed me to overcome my own loss.

SALLY REGENHARD: The strongest personal factor to me certainly, as with our other family members here today, was the powerful influence of my son. His life of integrity, of honesty, of compassion, of service to this country. . . my son was so unique. Besides being a graduate of the Bronx High School of Science and having a 146 IQ, he joined the Marine Corps in the family tradition, and served this country for five years before he joined the New York City Fire Department. He was such a compassionate person. He was such an intellectual person. He had such integrity and tolerance for all religions, for all people. I know that he could have understood the feelings of some people in Muslim countries and so on. He could see all sides of issues. His life, his integrity, his passion are what drive me forward. Like him, I feel have a higher responsibility; I want to help the public, I want to help save lives. He saved lives in his life and I want to continue that.

RICHARD SHADICK: What are the things that each of us can do as citizens to influence public policy to prevent similar catastrophes or improve our society?

SALLY REGENHARD: Well, if you live in New York, I think a fundamental thing that you could do to help influence public policy is to go out and buy a little book called the Green Book. In it is a listing of every single elected official in this city, their addresses where they work, their phone numbers, and their committees if they're in New York City government or state government. It also covers Congress. That book was the one thing that helped me start my organization. A wonderful community activist named Arthur Tobb, who is an organizer and a union person, told me about that book. When I was able to get that book I was able to find out who was responsible for what in this city, in this state, and in this country, and we just started writing letters. Make yourself aware of who your public officials are and then write them letters. Write letters to the *New York Times* and to other papers because elected officials hate letters.

TOM ROGER: It's funny, we had these questions in advance and what I wrote down was very similar to what Sally just said. I think, unfortunately, that there's a huge sense of cynicism about elected officials and public institutions and their responsiveness to us. I think those of us who have been knocking on doors and knocking on heads to wake people up on some of these issues do feel that if you do make the effort to try to demand better representation and responsiveness from your elected officials sometimes you can make things happen.

GENE STEURLE: I've been studying a period of time that theologians call the Axio Age. It's a period of time over two thousand years ago when peoples around the world all of a sudden came to realize that they weren't just there for themselves or their tribes but that they really needed to react compassionately to others and it's reflected in the Golden Rule whether it's by Confucius or Buddha or the Jewish Prophets or whatever. I've reflected a lot on that and I've come to the conclusion that advances in civilization came about not just for religious reasons but for very practical reasons. That, in our own lives there are a limited number of things we can do to other people that affect ourselves. For you students in the room, for instance, you think about how can I get this teacher to give me a good grade or change my life or how can I get somebody to like me or love me or if you're in our type of occupations how can I get a promotion or something like that.

And so what we really have is this strange aspect of life whereby there is only a limited amount of things we can do for ourselves. Yet, there's an extraordinary amount we can do for others. The terrorists realize that. They are trying to change governments, to change the way governments act. The terrorists recognize the power they have as individuals and they're using it in very mean, dastardly, and horrible ways. But they recognize the power they have as individuals. And I think that the people who, many centuries ago, discovered the Golden Rule came to the same conclusion: we do have extraordinary power to act. Just to give you a quote from Marian Pearl, who's the widow of Daniel Pearl, a *Wall Street Journal* correspondent who was killed by terrorists, not on 9/11 but in this same recent period of terrorism, said, "The task of changing a hate-filled world belongs to each one of us." And I really believe that. Each of us has a lot of ways we can act and when you go to our website we give a whole variety of things that people can do, including things for affecting government and writing your Congress person but we also include things like if you feel that you could work an extra year before you retire maybe you could build a school somewhere. And if you're a student and don't have much money well maybe you do a walkathon around the school building and raise some money for a cause.

So there are a lot of things that people can do and as I said we offer some here but the other panelists have offered a lot of others as well. I guess the main thing I would say is to encourage people to realize is that you do have tremendous power.

RICHARD SHADICK: Thank you. We'll do one more question and then open it up to the audience if they have some questions. One question that probably is on many of our minds, now that we are here in Lower Manhattan, has been quite controversial: What role, if any, do you feel our government officials should play in memorializing those lost on 9/11?

SALLY REGENHARD: Well, I think that the government, first of all, has to listen to the people. They have to listen to what people want and they have to be transparent, they have to be honest. They should not have another agenda of what business people and people who have economic interests want. They really need to listen to the families of the victims and unfortunately, in my opinion, I don't believe that was done. Government officials have to memorialize those lost on 9/11 by addressing the very painful, deadly mistakes of 9/11 and those of the aftermath, the procedures that were inadequate. For example, the reclamation of human remains from Ground Zero and from the buildings around the perimeter of Ground Zero. That was a failure in government, in city and state government. We still have hundreds and maybe even thousands of pieces of human remains scattered around the perimeter of Ground Zero. Two years after the City of New York signed off, for example, on the Deutsche Bank building, saying that it was free of remains . . . well, in the last several months we've discovered over seven hundred-fifty pieces of human remains. We can't have this denial of what is there. We need to do the right thing. Be honorable, have integrity, listen to the families who are crying out.

You know I'm one of the 42% of people who does not even have one single remnant of my son. The remains of 1152 people still are missing. They are surely around Ground Zero. They were not reclaimed in the proper way. We didn't have the right people coming in, the professionals, the right scientific organizations, to do that. It's questionable about what will happen.

That's what government needs to do. That is the best way to memorialize people. Do the right thing for their memory.

TOM ROGER: Going beyond some of the issues that Sally just raised, the actual memorial process that largely has been orchestrated by the Lower Manhattan Development Corporation obviously has been very controversial. I think they tried to be as open with the process as possible and by virtue of doing that involved many other constituencies besides the families and all the various conflicts between various people's priorities. I agree with Sally. In terms of what actually occurs in the form of a memorial, the families have to have a major voice. They aren't the only voice, unfortunately, because of where it's situated. And that's resulted in some of the delays and back-tracking on the process. Once the decisions are made and large-ly signed off by the constituencies, the government, however, needs to play a role to make things happen. Certainly the Port Authority and the City and the State have to play a major role in that process which is saddled with all the bureaucracy that comes with any large public project. I'm on the World Trade Center Memorial Foundation board and we're responsible for raising a large share of the money. We'll also have responsibil-ity for operating the memorial and the museum. I do under-stand that it's a difficult process, one not without frustration, but I think in the end there will be something there in which government can be proud of having a role.

GENE STEURLE: I think the main thing the government can do is to create living memorials to our loved ones. I'm not opposed to all the other memorials and things that we're doing but I real-ly want to honor their lives as best we can and what I mean by living memorials is not just things in the grand scheme like for-eign aid. For instance, there was this big fight in New York over a freedom center that was abandoned because of the location they were going to put it in—well, put it in some other location. Living memorials are things that would reach out and speak about freedom or speak about the wonderful sacrifices of these first responders, that inspire us to be something more. I've encouraged both the memorial group in New York and Washington to create little film clips on each of the individuals that were killed. That would be inspiring. They do this at the Holocaust Museum. They now try and give you a little bit of information about a person who was killed during the Holocaust and I think whether it's the benches in the Pentagon or the New York memorial we should be able to look online and get a little film clip about it. I'd love to know a little bit more about Sally's son. So I think the government really needs to think a little bit more about living memorials. And quite honestly, in many cases they wouldn't even necessarily be that expensive relative to some of the other things that are being considered.

SALLY REGENHARD: I absolutely agree with you on that. I'd love to see that.

AUDIENCE: My question concerns the person who has been described as America's Mayor and who may be elected president in part because of the way that the world views him in relation to 9/11. There's a new book out, *Grand Illusion Rudy Giuliani*, that argues that his policies exacerbated loss of life during 9/11. And my question is, where do you see Rudy Giuliani; does he deserve the hero label that he has come to receive; or is this book accurate?

SALLY REGENHARD: As I mentioned, 9/11 was characterized by many failures, at least here in New York City. There was a failure in that we did not have an integrated command structure whereby the police department and the fire department could communicate with each other. We had no communication between two of the most critical agencies necessary to save lives. So even though the police department was able to see that the second tower was in great danger of collapsing—they were able to see that 20 minutes before the building collapsed—my son and the other three hundred forty-four firefighters had no way of getting that information because the administration in power allowed a culture of rivalry to prosper. New York City should have been prepared for terrorism. We were hit in 1993— 1993 was a microcosm of 2001. Nothing was learned by that administration. We had a failure of emergency planning and we had a disaster in emergency management. If I knew that my son would have been given the same radios that failed in the World Trade Center in 1993 I never would have encouraged him and supported him in his decision to join the Fire Department. He joined the Fire Department because he wanted to help people. It was something that he could do while he finished his studies in art and in writing that he had started in California. A book like that, that you mentioned, *Grand Illusion*, it's the

126

first bit of truth that we've had about the failures and the deadly mistakes of 9/11. We have to look at these issues because if we are ignorant of deadly mistakes of the past we will be doomed to repeat them in the future.

AUDIENCE: I want to say how inspiring and awesome all of you are. I'm filled with awe, with respect, to all of you on the panel for dealing with and coping with the devastation in the way you have.

I've served as a consultant to the Fire Department as a psychologist and have worked in trauma since 9/11 and it's very hard to get a picture of how all the victims and their families and the people around us have been doing. It's often not as rosy as we see in the media. So, I wanted to ask you, if you could, to reflect on the fate of victims' widows and children that you know of and comment on the backlash I've been seeing for the psychic dimension of the victims' lives lately. If you read *New York Magazine* this past month they had another article blaming the victims for the stoppage of work in the World Trade Center. I just wanted to see if you shared that view or had any thoughts about that, so it's really a two part question. Thank you so much.

TOM ROGER: Certainly many of the families and those directly or even indirectly affected by the loss of life are still suffering quite substantially. Clearly, for those of us that do our best to try to think about how our loved ones lived and not how they died, this time of year makes that really difficult. I know my wife always says that September is the worst time of year for her. So clearly don't take our persona as representative of all the people out there. There are a lot of people that are hurting.

Gene's organization does a wonderful job at giving lots of different places for family people to get involved in whatever way that suits them and to give them a positive outlook. And I think that's a wonderful thing. As I know a lot of family members, I think that's where they struggle. They felt so helpless on 9/11 and they felt so helpless and abandoned afterward and they really didn't know which way to go and I think that we need more of those kinds of organizations.

SALLY REGENHARD: I agree with Tom on this and certainly about the wonderful work of your organization also.

I want to say that the parents are a large group that has been somewhat overlooked. Certainly the widows and the children are very, very devastated. But for the parents, I must tell you based on my experience with my other group, 9/11 Parents and Families of Firefighters and World Trade Center Victims, the parents are doing terribly. And we may be up here as parents and loved ones, and as Tom very aptly said, we may look as if we have it all together, as is if everything is good, but it's not, it's terrible. My husband and I cry every day for our son. The parents of the victims have aged decades in the last five years. Their physical health has deteriorated along with their emotional health. We're doing very, very badly. There's nothing worse than knowing that the promise of a young life has been lost. The average age of a World Trade Center victim was only thirty years old, thirty years old. We have lost the hope of a future and our beautiful children.

Finally, I'd like to mention what you said about blaming the victims. There is a lot of blaming the victims and that's very unfortunate. Some of the victims who were blamed were the New York City firefighters. People from the former administration actually had the audacity to testify to the 9/11 Commission right here in New York City that the firefighters heard the orders to evacuate yet decided to stay and die. That is a lie. That is an abomination. My son and these other beautiful young people, whether they were firefighters or civilians . . . who on earth would have chosen to die in that building for nothing? So yes, there are a lot of inequities, there's a lot of blaming the victim. But, people who are running these organizations were working for different reasons. I know that we're all united here and we all have the same goals. God bless all the people who died and really God bless all the families of the victims because it's a devastating role to have, but in honor of our loved ones we're going to continue with our work with the help of God.

RICHARD SHADICK: I want to thank each of the panelists for coming and sharing their personal experiences. I think they showed a really a great example of the resilience and growth that I referenced earlier. The pain does not go away; that's true. But we do see that there are individuals, many individuals, who can make some good out of the disaster that took place five years ago. Thank you all for coming.

GENE STEUERLE: Richard, could we each again mention our organizations just in case people want to get in touch with us?

For us, it's *ourvoicestogether.org*, and for those of you in the audience there are some sign-up sheets outside. Nothing more than just getting you on an email list. We can send you a lot of material. And maybe the two of you would like to mention how to get in touch with you as well.

SALLY REGENHARD: My organization is *skyscrapersafety.org* and a website we created for my son *christianregenhard.com*.

TOM ROGER: *familiesofseptember11.org*.

PANEL SIX
REBUILDING, REPAIR, AND HOPE

KATHRYN WYLDE: Good afternoon. I'm Kathy Wylde from The Partnership for New York City. I would like to thank Pace University and Dr. Caputo and Meghan French for bringing us together and giving us this opportunity to reflect upon the fifth anniversary of 9/11 and, importantly, to think about the future. I hope that our panel today will focus in particular on the future and the opportunities and challenges ahead of us. The members of this panel all represent organizations that have been actively involved in either the private or public sector mobilization of the response of both New York City and State to the terrorist attack, and the organization of the recovery and rebuilding efforts. All have a real stake and interest in the activities of the last few years. I'm going to ask them to introduce themselves and say a few words about the role that they and their organizations have played in that process. I think before I do that, though, I'm going to ask John Cahill, our panel opener, to say a few words, summarizing the status of where things are, and where they're going, in terms of the state's leadership of the rebuilding effort. John is the Secretary to Governor Pataki, and has been a spirited and wonderful force in leading the state's efforts to bring together all the parties that have to participate in financing, rebuilding, and planning the redevelopment of the Trade Center site and the renewal of Lower Manhattan. He spearheads that effort on behalf of Governor Pataki. He has been a terrific person to work with and has done a great job.

JOHN CAHILL: I want to extend my thanks and gratitude to Pace University and to Dr. Caputo and Cindy Rubino. I am a proud two-time graduate of Pace University, and it makes me proud to see what Pace has done for this community, particularly since September 11. They've really been stalwart in helping with the revitalization of Lower Manhattan, so thank you, Pace University. If I could, I'd just like to take a couple of moments,

131

before we talk about where we're going with Ground Zero in Lower Manhattan, to really refresh people's recollection about where we've been since that fateful day five years ago. And I think, when you look at what's been achieved over those past five years, you'll agree that the progress has been remarkable. Yes, there have been setbacks, and maybe the setbacks are to be expected. But when you consider the enormity of the damage of September 11–having lost the lives of 2,749 New Yorkers and people from around the world, having displaced fifty-five thousand people, having sixty thousand people instantaneously unemployed, having lost thirty million square feet of office space—it really is extraordinary how far we've come. Since September 11, 2001, Governor Pataki has put enormous focus on public participation in the redevelopment of downtown. I'm sure that many of you who are here today have been here before, to speak out on the issues that concern the redevelopment of Lower Manhattan. In 2002, when we had the formation of the Lower Manhattan Development Corporation (LMDC), five thousand people showed up at the Javits Center to speak out about what should happen in the redevelopment of Lower Manhattan. The LMDC received an unprecedented 25 million hits on the website about what should be done in Lower Manhattan. From there we had the selection of Daniel Libeskind's master plan, the focal point of which was the memorial. That, obviously, continues to be the focal point in our work with respect to Lower Manhattan. Around the memorial will be ten million feet of office space, topped off with the Freedom Tower which will send a signal to the world that, yes, the skyline of New York is important and we are going to build bigger and stronger than ever. We already have had the unveiling of the signature train station, which will be the epicenter of economic activity in Lower Manhattan. And since that time, in a worldwide competition that elicited responses from 5,200 people from around the world, we've had the selection of the memorial by Michael Arad. The process that the governor and the mayor have really tried to hold true to is a public process, because really what was attacked on September 11, very much, was our democratic values. The governor has always felt that it's very important that

132

the new design of Lower Manhattan not be just one man's stamp. Each and every American is going to have the opportunity to say how this city should actually be rebuilt. And as you look out today, five years later, we have every significant component of the Master Plan underway. The memorial started excavations for the laying of the foundation two weeks ago, so the Freedom Tower is underway. The PATH station, too, is underway. It's also important to look not just at Ground Zero. Yes, that is the heart of our efforts, but it's more than that. When we undertook this task, we understood it was more than just about the site, it was about Lower Manhattan. That's when Mayor Bloomberg came out with his vision for Lower Manhattan, the river-to-river vision for this wonderful part of New York City. And since that time the construction for the Promenade on the south side has been completed. The north side is under final design, and construction will commence there some time next year. We have the new transit hub for the MTA, which is going to untangle the web of commuter lines into Lower Manhattan, completing what will be the finest transportation infrastructure of anywhere in the world. So, here we are now, truly poised to move forward with the real construction of redevelopment. I often use the analogy that this is like painting your house, or painting a room. The hardest part is doing the preparation work and we're near completion of that. Right now we have about four hundred individuals working at Ground Zero. That's going to increase to twelve hundred a year from now, and twenty-four hundred a year after that. Lower Manhattan will receive about $30 billion of investment; there is an unprecedented construction interest in Lower Manhattan. I think that's because people see the confidence of everybody who has been involved and what they have achieved. Whether it's Goldman Sachs, whether it's Tiffany, whether it's the hottest residential market in New York City, people are speaking with confidence about downtown. And yes, there have been setbacks along the way, the Freedom Tower obviously the most notable one with the concern about security issues. But we didn't quit. We went back. David Childs did a magnificent job coming up with a new design for the Freedom Tower, a design that I think actually works better for the site.

The trademark over the last five years has been perseverance, it's been public commitment, it's been spirited debate. Yes, there's been some criticism. But all that is a part of the democratic process, which is so important, at the end of the day, to a successful redevelopment in Lower Manhattan.

Thank you for your interest in Lower Manhattan

KATHRYN WYLDE: Thank you, John. I would like next to introduce Stefan Pryor, who is President of the Lower Manhattan Development Corporation for another week.

We should tell our audience that in a couple of days, Stefan goes over to join Cory Booker and become Deputy Mayor of Newark. Stefan was formerly with The Partnership for New York City, the organization that I'm with, running our education programs.

STEFAN PRYOR: I'm the outgoing president of the LMDC. I have had the true honor of having served with the LMDC, coming over from the Partnership, as soon as it was formed by the governor and Mayor Giuliani. If I could stray from the script a little, I would like to thank John Cahill, a man whom I have worked with all these years. You can tell from his remarks that his dedication and knowledge are profound. To be part of his team has been a terrific honor. I also would like to thank the governor and mayor for having appointed me to the position of head of LMDC. I will serve through September 11, the fifth anniversary, and then I do move to Newark. My heart, though, will remain here. I am a New Yorker, born and raised, and look forward to monitoring and assisting in any way possible going forward.

KATHRYN WYLDE: Would you like to say something about the LMDC's role? What they've done, and where they're coming from?

STEFAN PRYOR: The LMDC was created by the governor and the mayor, in 2001, as the coordinating entity for the rebuilding. The LMDC sets forth the plans and the financial strategies for the rebuilding, which is to say that in the earliest phase of the rebuilding we set forth principles on a blueprint that articulated the basic tenets by which the rebuilding would be conducted. As John Cahill described, having an open, public process turned out to be one of the hallmarks of the process. The decisions which have come from this process include having a memorial as the centerpiece of our work, ensuring that the street grid at the World Trade Center site would be restored to its previous condition, selecting a Master Plan, and launching additional studies and competitions for other work happening throughout Lower Manhattan. When you look at the Fulton Corridor plan—the extension of Fulton Street beyond the World Trade Center site in a revitalization effort that stretches river to river—that, too, is an LMDC effort. Of course, jumping back to some of the points I've already made on the memorial itself, the LMDC obviously conducted the competition that led to the juried selection of Michael Arad's design, and has served as client for the creation of the memorial, directly working with the memorial architects. We're now transitioning that responsibility to the World Trade Center Memorial Foundation. The Port Authority has, in a terrific move, agreed to, with enthusiasm, build the memorial. So, those are just a few of the things the LMDC has done. I should point out that Bill Rudin is one of our prized Board members.

KATHRYN WYLDE: That's right. Bill Rudin is a major property owner in Lower Manhattan, and the city as a whole. He has led the revitalization of Lower Manhattan since the early 1990's. Now, he has to turn around and do it again. Bill currently is the head of the Association for a Better New York, and his observations, both as a civic leader and a real estate owner, are invaluable.

WILLIAM RUDIN: I do wear many hats. In addition to what Kathy mentioned, I was a [founding] member, along with

Charlie Maikish, of the creation of the Alliance for Downtown, the business improvement district, where I still serve on the executive committee, and as a board member. I am on the board of The Partnership for New York City and president of my family real estate company in New York City. So, I come at the revitalization of New York from many different viewpoints, but with one fundamental philosophy. My family has been investing in Lower Manhattan since the 1950s. In the early 1990s, Lower Manhattan had over thirty million feet of vacant space. We believed that it had to be re-engineered to be dynamically made into a 24/7 live-work community. And the mayor and the governor, and, I think, everybody involved in Lower Manhattan, agreed with that point. We need institutions like Pace; we need cultural institutions; we need open spaces like Battery Park, where Castle Clinton is. And we need to intertwine all these important organizations and institutions to work together. That is the goal that the mayor and the governor articulated right after 9/11. We have the opportunity, now, to look at Downtown in kind of a different way, to do things that we were hoping to do, but couldn't afford. We now have the Federal money and resources to be able to create a master plan for transportation, to try to get a railway from Lower Manhattan to JFK, to have more open space, and better public transportation, things like that which are important to attract both the commercial tenants, as well as the residential tenants. And you're seeing the fruits of that labor with the tremendous influx of more residents. We've been talking about diversification, away from financial service industries, as long as I've been in the real estate business, and today it's happening. In the last eighteen months, almost two million feet of office space has been leased by companies who had never before been in Lower Manhattan. They are diverse: they are media companies, they are advertising agencies, they're design companies, they're international companies, which is I think a very powerful statement. That's a very important statistic and it's something to keep in mind as we move forward, and read the articles which question whether or not Lower Manhattan is coming back. The answer is, 1000%,

136

it is. So we're very optimistic. I'm looking forward to working with everybody as we move forward.

KATHRYN WYLDE: Thank you, Bill. Stefan and Bill referred to the amazing public mining process. One person who was very much involved in that process from the private non-profit side is Bob Yaro, president of the Regional Plan Association (RPA). Bob has been a force in organizing community and civic involvement in the efforts to recover and rebuild.

BOB YARO: The few and the proud. I think I met Stefan Pryor the first day he was on the job at LMDC, and I'm going to get to work with him on his last day. RPA is a regional group in Lower Manhattan, but we're also working with Stefan and Corey Booker on the new master plan for Newark, the second largest city in the region. After 9/11, we were very shocked and dismayed. Like everybody else in New York, we felt committed not only to rebuilding this place, but to trying to make Lower Manhattan better. We had eighty-five civic groups working together towards this. The idea was not only to create a vision for what we call the first Twenty-First Century City in sustainability, in accessibility, and livability, but also to make sure that the public had a strong say. There was so much emotion after 9/11, and emotion continues to be wrapped up in this process. This is not a conventional rebuilding process. John, you've been in the eye of the hurricane, right? You've seen the emotions, we all have. What we try to do is channel all of that emotional energy into a constructive process, to work with the LMDC and with the city, and the other public authorities to come up with a rebuilding plan that has broad public support. That's what's notable about this. Every other major rebuilding process, at least in my experience in New York over the last generation or so, was that it always got mired down in controversy and delay and litigation. And, yes, we've taken steps forward and steps backward, there have been re-appraisals and so forth, but we have a strong commitment to move forward. The engagement of the civic community throughout the process has been a very important, and constructive part of it.

KATHRYN WYLDE: I think you can tell that all of us, everybody on the panel, are extremely enthusiastic about where we are in the process of planning and re-development. In terms of the general public, there has been a lot of concern, a lot of criticism, a lot of skepticism. I would like to give the panelists the opportunity to talk a little bit about why there is a gap, about why it is so difficult for people to feel the same positive feeling that Deputy Mayor Doctoroff expressed yesterday. He said, "You know we're at the midway point, we're at the five year point in a ten year plan, we're halfway there, and things are going great." And you could tell that people around the room, who weren't that involved were saying, "Well, yes, that's the glass half-full version." But on the other hand a lot of people see the glass half empty, too. So, is it communications? Is it politics? Is it that people don't understand how long it takes to get anything built in New York City?

JOHN CAHILL: I think one of the issues is trying to convey the complexity of trying to rebuild a sixteen acre site. To put it in perspective, it's about the same as rebuilding all the commercial space in the city of Baltimore on sixteen acres, eight of which are going to be occupied by a memorial. Yes, there are public expectations. Every American, every New Yorker wants to see this site redeveloped as quickly as possible, and that's only normal. It's a scar on our city, and our country. We want to see that wound healed. We are all anxious to move forward. The most important thing here, is that we move forward appropriately, thoughtfully, doing appropriate planning. Now that we are ready to construct, the construction can move expeditiously. It is difficult to convey the complexity of what's being undertaken by the Port Authority, by the MTA, by the State DOT, by the city DOT. Throw in the security concerns of the NYPD and the Port Authority—the job is immensely complex. While we strive for consensus, we're not going to make everybody happy in this process. What we have tried to do is provide platforms for people to express their concerns, and when you have that type of energy, a lot of that energy is going to be released in terms of frustration. I think when people look back at this process ten to twenty years from now I'm very confident they're going to say that we did it properly, and frankly that we did it pretty expeditiously. It took, I think, 28 years to build the first World Trade Center. A lot of the frustration comes from the difficulty in communication, a lot of it is understanding, and, yes, the process has been thrown into the political mix, as many things do here in New York. All of that is understandable, and frankly those of us that are in this can't really gripe about it, because that's the world that we all live in. We should be willing to accept those criticisms, those dynamics, of the challenge.

STEFAN PRYOR: Picking up on John's point in regards to the complexity of the process and the multiple voices within the public in regards to the rebuilding, I think there is a tremendous sense that we all share in rebuilding, and to move forward expeditiously. I think we're all convinced that we are doing that effectively. We're proud of where we are. But the fact of the matter is that within the set of engaged constituencies there are different points of view on what we ought to do. To this day, we get correspondence from people saying, "Rebuild the Twin Towers." We get correspondence from people saying, "It's a sacred site that ought not to be rebuilt at all." We get correspondence in regards to the more nuanced sub-components of the project: how the names of the victims ought to be organized in the memorial; the precise design of the memorial; the array of towers at the World Trade Center site; whether there ought to be residential structures; where it ought to be in the mix of development downtown. These are legitimate questions for people to discuss. And the fact of the matter is that all of us on the dais who have been part of the story, question what the right path ought to be. I think that we have taken the right amount of time, thanks to the leadership of people like Bob Yaro and the civic organizations, to take into account public input. It is now time for us to be building. And you know something? We are. If you look at every corner of the sixteen acre World Trade Center site, there's construction underway. It's quite remarkable. The Freedom Tower, the memorial itself, the PATH station, all of these projects are under way. Our initial plans that we presented in July 2002 in the Javits Center — people hated those. Unarguably, it was part a communications failure. We learned a lot from that, including a set of imperatives as to planning. One of these imperatives was to build a promenade along West Street that would be compatible with the memorial that we would build, a promenade that would help stitch together Lower Manhattan. Well, you know what? The lower promenade isn't just a concept, it's being built. And the upper section will be in construction soon. A tremendous amount of construction is already under way. And I do think that, as John's articulated, as Dan Doctoroff has articulated, when we look back from a few years hence, we will say, "What we have created is magnificent, and it is miraculous that we've done it within this compact area between two rivers in such a short period of time."

KATHRYN WYLDE: The other thing that I don't know that people really realize, is that while the planning process on the site is a public planning process, the insurance process only paid for a relatively modest portion of the total development costs. It rebuilds the trains and the infrastructure, improves the public amenities. But we have to be building and designing to attract private capital investment. I think that's also a disconnect. It's not as if the government is sitting there with a piggy bank. We aren't in that situation. We have to design for a market that is always uncertain.

WILLIAM RUDIN: The competition with Midtown to attract tenants is growing. The market's very expensive in midtown, and we want to keep attracting companies to New York, to keep them here, to having them grow. We need first class, high technology, big floor plan buildings, and we have the opportunity to create them here. Even in the early sessions, there was no debate about ten million feet of space. I think we have a plan, today, that makes sense. The other point is that we shouldn't be looking just at Ground Zero. It's obviously a very important aspect, a significant aspect, but as the mayor has talked about and the governor's talked about, we have to be looking at the whole of Lower Manhattan and what is happening in other parts of the downtown area. You walk the streets today and you see cranes, you see apartment buildings that weren't there five years ago, you see people living and taking their children to parks that weren't in existence five years ago, you see new restaurants, you see new retail. Do we need more? Absolutely. But if you think back to September 12, 2001, and ask where we would be on, September 11, 2006 . . . I don't think anybody could have imagined the progress we've made today.

BOB YARO: The reason, of course, that people have said that there hasn't been any progress, like when New Orleans mayor, Ray Nagin, put his foot in it complaining about our hole in the ground, is that there is still an open sore down there. People can't see the amount of effort that's gone into rebuilding the infrastructure of Lower Manhattan. Rebuilding not just the sixteen acre site, but the other square mile that we set out to transform. You must look at the investments that *are* in the ground, that *are* under construction now: transportation, new infrastructure, the reconstruction of virtually every street, all new utilities in the oldest business district in the United States. A lot of people don't put it together. They don't ask, "Why is all this construction going on here at Fulton Street, and what does this have to do with the bathtub?" We've made a lot of progress in the last five years. We've probably made as much progress over the past year, as we have in the previous four years. A lot of it is because the foundation's already been laid. Even in the first regional plans drawn up in the 1920s, we were talking about needing to bolster the business sectors in Lower Manhattan, which were even then moving to Midtown. What we're doing here is creating a new foundation for the next century of economic growth in Lower Manhattan with investments in infrastructure, with ten million square feet of new commercial space on the site, and another several million square feet nearby, with Goldman and the others. It's hard for people, for casual observers, to see this all together. We did take a step back, but in the end we created this foundation that's going to make it possible for Lower Manhattan, maybe for the first time, to really go forward with confidence in its role as the second largest business district in the United States.

KATHRYN WYLDE: In the last couple of weeks we've been reminded of the anniversary of another disaster, a natural disaster, Hurricane Katrina. It made me think about the challenges we face in this country in dealing with all types of disaster. Our experience in New York showed that we weren't well prepared in terms of Federal programs and policies. I think people forget that we spent the first two years after 9/11 trying to pass some Federal laws, and reframe some Federal programs that originally were designed for floods in rural America. Do any of you have thoughts about where we are, right now, in terms of our policies and programs? It seems like many of the criticisms of the first year after Katrina, of the state, of the city, and of the Federal Government and FEMA, echo much of what we hear here. What else should we be doing, and what should the Federal Government be doing at this point?

BOB YARO: You know, one thought is that we understand in a better way just how important it is to have resiliency and redundancy built into big infrastructure systems. A remarkable thing happened here. We forget that there were, on 9/11, 14 subway lines in Lower Manhattan, as well as the PATH system. We lost several of those subway lines that were damaged on the Trade Center site. And yet, a week later, Lower Manhattan was back in business. The transit system was up and running. The N and R—the tracks that were destroyed when the towers came down—were back in service eleven months, twelve months later. The PATH system was back about two years later. It's a remarkable story. If we had had only had one transit line in Lower Manhattan, like we did for the PATH line we lost, we would have been out of business. But we didn't. We had fourteen subway lines, and ten of them were not damaged. It underscores the importance of creating redundancy and capacity in our transportation system. In most of the United States, it wouldn't take a terrorist attack to shut down the infrastructure, it only takes a heavy dew, or a rainstorm, or an overturned tractor trailer, because there's no capacity. I think this is an issue for us as we think about trans-Hudson transportation and the Second Avenue line, which would give us redundancy on the East Side.

KATHRYN WYLDE: Even the need for having competitive ferry systems, to get people here, is one of the things that came out of the experience of 9/11, and put some competition in the harbor.

BOB YARO: That's right, so we have an unfinished agenda. Though a lot has happened. Second Avenue, the East Side, these other transportation projects give us a lot of redundancy that we didn't have before.

KATHRYN WYLDE: Stefan, I know you did some work with New Orleans. Can you tell us a little about that?

STEFAN PRYOR: One of the things we did in the Lower Manhattan recovery process was create a system for dealing with this level of destruction and developing a rebuilding process that was unprecedented. The fact of the matter is that our money at the LMDC flows from a community development program at HUD that's one of the most flexible programs in the Federal Government. It typically is utilized in urban centers for conventional revitalization projects, not after disasters. HUD re-crafted that program through Congressional legislation to be even more flexible, and to operate differently. That is the precisely the formula that was ultimately given to Louisiana and the Gulf Coast states. So one of the things that we did, the governor's office did, and the city did, is to help the state of Louisiana set up the kind of systems that they needed. They created, ultimately, the Louisiana Recovery Authority (LRA) which, in many ways, parallels the LMDC in the way it's going to behave and operate relative to HUD. One of the things that's very critical for them is that they've already suffered scandals in the rebuilding effort. One of the areas in which we've been very fortunate is that we've had such a clean effort here. It was not by chance, it was very deliberate on our part. Already, after

143

Rita and Katrina, where there was some unfortunate activity, the LRA is suffering in that aspect. They have to operate clean. This is one of the things, I think, that we can help them with. The HUD Inspector General recently stated that he thought that the LMDC was the model for the nation. Our multiple layers of protections as to how funds are used are why we haven't had scandal here. It's been a clean operation.

KATHRYN WYLDE: That point has not been made enough. To have this multi-billion dollar undertaking with so many players receiving grants and funds, the scandal free nature of the experience is something to be very proud of.

STEFAN PRYOR: There are multiple features of our integrity structure which we think are irreproachable. Simple things. The governor appointed Frank Zarb to our board and said, "Let's make this the tightest ship ever seen." Frank said, "Fine. Let's have private sector style protections here. We are going to hire an investigations department. We're not going to wait for the Attorney General or the District Attorney to uncover fraud; we are going to bring our cases to them." And that's what we've done. We have actually detected cases of residential grant program abuse and brought them to the law enforcement authorities. That's exactly what ought to happen in other disaster scenarios. We have an internal auditor and an external auditor; we have a board committee at our agency that directly oversees the audit process. Not to belabor the point, but there are many replicable elements of what we've done in concert with all these law enforcement agencies who've been terrific, every single one of them works to ensure that we have this clean process. That kind of deliberateness is what will get others through it.

KATHRYN WYLDE: John, do you empathize with Governor Blanco and what she went through in terms of dealing with Washington and with FEMA?

JOHN CAHILL: Certainly. I had many conversations with her chief-of-staff about how many times they were trying to fit a round peg into a square hole. I haven't been to New Orleans, but I really hope they are in the same place four-and-and-half years from now that we are: having a comprehensive land use plan, and coming up with visions for the future of the city. Stefan's point is well taken: it takes strong acts of leadership, and not finger-pointing, to make it happen. I really hope that New Orleans can get to where we are.

KATHRYN WYLDE: It is important to recognize that everything is good here, good but not perfect. *Crain's,* this week, ran an article talking about some of the struggling small retailers whose customer base has not come back, yet who are feeling rent pressures as property values rise in the neighborhood. Sometimes the good news and bad news get mixed up. I would like each of you to comment on what your biggest concerns are, with the challenges that are out there, things that you're worried about getting done, whether it's fundraising for the memorial, or whatever. Bob, we'll start with you.

BOB YARO: Well, it's a neighborhood in transition, and I think we just need to see the process through. As everybody has said, it's not just about the site itself, it's making sure, as we move forward, that we have design guidelines for the public spaces, the public realms. That the final plans for the memorial and the new street grid and the other public spaces—whether they're designed by the MTA, or by the Port Authority, or by the City of New York—are coherent and legible. And there must be some kind of continuing coordination. If you look at the streets of Lower Manhattan, there's so much reconstruction going on. And yet I don't get the sense that we've really locked in on a larger public realm strategy. After the IRA bombings in London and Birmingham, Central London transformed itself into a largely pedestrian precinct for security reasons. It's added a tremendous amount to the amenity and quality of life of the district. We ought to be considering it, yet we still haven't finished the transportation management plan five years later to make decisions about how to use the limited curb space, how to handle deliveries, and security concerns—something other than tank traps. That's an agenda that should be considered here. Let me say one other thing, when we started this, people asked me how long it was going to take, and I said it could take a decade to get this site restored. It's going to take a generation to transform Lower Manhattan. As Dan [Doctoroff] said yesterday, we're on track, but, and this should come as no surprise, there are going to be business cycles and there are going to be real estate cycles that we must go through to move forward. I think we must continue to focus on these coordinating functions, to make sure the follow-through is there, so that the political change goes on and that there is follow-through. I think that's something really to keep our eye on, because it's going to take the rest of the decade to finish this job.

WILLIAM RUDIN: I agree with Bob. Several of our buildings have still not really recovered, in terms of the market moving up, because they're older buildings, they were built in the 60s. These are office buildings. They were impacted by [new security provisions in effect around] the stock exchange and the new streetscape that hasn't been completed yet.

What Stefan has been working on, and what we're trying to push for, is an appropriate redesign of Broad Street and Wall Street that not only is user friendly and doesn't scare people away, but that provides the security that the Stock Exchange needs. As well, very soon you'll see residents moving in, literally right across the street from the Stock Exchange. The master plan has to be funded and built in an expeditious way, and a lot of it's already done. But a lot of things still are in the planning stages. So, we must keep our eye on the ball and keep pushing where we think it's appropriate.

STEFAN PRYOR: I think one of the greatest challenges is going to be the amount of activity that's happening in one place. The reality of the situation is that very soon the traffic congestion, the noise, and the perceived and real disruption that will occur because all the construction that's happening downtown is going to become a challenge. A switch is going to flick on in people's minds very soon, with the amount of work that is already under

146

way, and the amount of work that John Cahill's coordinating, and that Charlie Maikish is working on every day up at the Command Center, that this is going to become the reality in Lower Manhattan. I live downtown and work on Liberty Plaza, and I can tell you that there is construction day and night. It's music to me; I think it is the most beautiful music in the world. But it's going to force people to pay attention, whether in terms of navigating the congested streets and changing or monitoring the closings of some of the rail systems. All of those things are terrific, but they are frustrating, too. To the degree that there is frustration, I hope that it will transform into enthusiasm for what is being done. I hope it allows people to be flexible.

JOHN CAHILL: Stefan hit on what is going to be the next big issue and that is managing the construction that is going to be happening in Lower Manhattan. It is going to be of a magnitude that probably has never been seen before. The challenges facing Charlie Maikish's group, the city planning commission, the city DOT, the state DOT, and the residences and businesses are going to be enormous. Everybody's already spent a lot of time planning, because that day is really, literally, around the corner. From week to week, as construction activity has grown in the pit, it's also been growing outside of the pit, in Lower Manhattan. It's imperative that we manage the environmental issues that are going to be related to the site, to truck traffic, to the diesel trucks. The community, rightfully, is going to be concerned—they need to be as involved in the process as they have. Businesses are going to be affected. Smaller businesses, Kathy, to your point, will probably be the ones further hurt by this kind of activity. We're all cognizant of this. These are the issues that I see are going to be the most pressing. The construction within the site is going to be of enormous magnitude. How this affects the external community is going to be of greatest concern.

KATHRYN WYLDE: My office is five floors above the South Ferry project where they're extending the Subway. There are explosions which shake the building every day. It's really distracting. On the one hand, you're so happy that it's going to take seven to eight minutes off everybody's trip to Lower Manhattan from the West Side, and that you won't have to ride in the front four cars any more—the long range benefits are terrific. But the short-term disruption of both the streets, and of doing business is tough and it's just a taste of what's to come. Obviously the biggest concern that everybody has is preventing another terrorist attack or incident in New York, and a huge amount of resources are going into that. I can't leave the conversation without mentioning the concern of the business community about the Homeland Security allocation of federal funds. Despite the efforts of the Governor and the Mayor and all of us, to fight for an allocation based on risk—New York City has 50% of the nation's insured risk—we get a very small proportion of Homeland Security funds to fight, and prevent a future attack. This not only is the top of my concern, it is the same for a number of us.

JOHN CAHILL: Kathy, you're absolutely right. Whether the issue is the construction vehicles on site or developing security measures within the confines of constructive activities, or even planning for the type of security measure the site will have once it's complete—the effort will be unprecedented and it will be expensive. It is necessary that the Federal Government and Homeland Security come to the table and basically fund those security measures—unprecedented security measures—not just for the site, but for all of Lower Manhattan. Lower Manhattan is the financial capital of the world. It is also the site of two previous terror attacks. There has got to be risk-based funding. Hopefully the administration and Congress will not wait until some other unfortunate event. Our top priority must be, now and in the future as we develop the site, to secure federal Homeland Security funding, because that burden should not be borne solely by the city and state.

KATHRYN WYLDE: I'm going to open up the discussion for a few minutes for questions from the audience.

AUDIENCE: I'm the associate university librarian here at Pace University, and it's very exciting to hear about all the wonderful, transformative efforts that are being undertaken for Lower Manhattan. It is going to be very exciting to see what happens over the next few years. My question relates to Pace University. We are probably the major university in the area. President Caputo has very wisely placed us at the table of the planning process through the Center for Downtown New York. I would like to get a sense as to what you feel are the contributions that we can be making to the development of Lower Manhattan, and what advantages we might gain from a transformed Lower Manhattan?

JOHN CAHILL: Since September 12, 2001, Pace University, as the largest institution of higher learning in this vicinity, has been critical, not only in their response and help in the recovery effort, but in educating and participating in public forums with respect to the development of Lower Manhattan. President Caputo and I have had numerous conversations about what's necessary to push the agenda for Lower Manhattan. Pace has always looked to be in the forefront, partnering with us, looking for opportunities, advising, frankly, me, and I'm sure the LMDC, and attending to the concerns that the students have, that the faculty have, and that the residents have. Consider what it's been doing this week. Pace is providing not only a critical forum for a retrospective on where we are five years after 9/11, but the opportunity as well to look forward to the redevelopment of Lower Manhattan, to come up with thoughtful ideas on how we can do things better, as we move forward now into this next stage with regard to the redevelopment and construction activities. Pace plays an important role in helping us to solve the very, very difficult questions about developing Lower Manhattan and managing and mitigating the impact that it has on businesses and communities here. As I said from the beginning, I really do feel proud of having been a Pace graduate, but I'm even more proud of what they've done for Lower Manhattan since September 11.

AUDIENCE: I'm really proud that we are the financial capital of the world, really proud of Goldman Sachs's vote of confidence. But I'm a little nervous about how much residential construction is going on downtown with regard to tax bases and unemployment. My question for Mr. Rudin is, how do you recommend we make a good mix?

WILLIAM RUDIN: That's a very good question. It's something that's been discussed at many different levels of the Alliance, with the city and the state, and Speaker Sheldon Silver. We forgot to mention that this is the district of the Speaker of the New York Assembly, Sheldon Silver. It's critical that we ensure that we have the right balance. In midtown, you have an office building and right behind it you have an apartment building. That's a wonderful mixture, and that's what makes midtown so very powerful, and such an attraction, that mixture of all the residents, commercial, cultural, restaurants, services. The more residents you have, the more retail comes. We've had a very difficult time, over the last ten or fifteen years, attracting retail downtown. It was always part of the debate about how you get people to live down here. One of the answers was that they aren't coming down because there's no services, there's no retail. But the retailers are coming. Even though some of those that are here now are having problems, we're now attracting some very significant names. And that will attract other retailers. And there is a concern, from some businesses, about the retail and the residential component. But there is a balance, and market forces will dictate it. You know, a few years ago everything was getting diverted to residential. As the commercial market gets stronger, however, maybe some of those buildings won't get converted. Plus the incentives for conversion from office to residential went away in July. So that will also slow down that process. You asked an important question.

KATHRYN WYLDE: Lloyd Blankfein, the new CEO of Goldman Sachs, is one of our board members. I asked him about his experience now in Lower Manhattan—you know Goldman has made a huge commitment to build two million square feet of additional space down here, to maintain their headquarters and 9000 employees—and he said, "You know, I walk out the front door of my building on Broad street and what I see going by is baby carriages." He's been down here for thirty years, and this is a new experience. So I think the mix is working. It's a whole new thing for Lower Manhattan, but I think it's working in terms of how people who work here feel about the neighborhood, how they really feel it's part of a community, that they're leaving at midnight and there are people around. I think there are real advantages to the residential mix coming in. There was concern expressed by some of the businesses downtown, particularly over in the World Financial Center, that they not be overwhelmed by residential. And early on, when there was concern about whether there would be commercial tenants and there was talk about building residential on the World Trade Center site, there was a lot of concern that the balance was going to shift. But I think people now feel that it's appropriate that the older, obsolescent office buildings are being converted to residential. The new buildings that are being built now are committed to be basically commercial. That's really given people, at least it's given the business leadership, the confidence that they're not going to be stuck in an island of residential. They feel very good about it.

STEFAN PRYOR: We fought very hard to make sure that there wasn't residential on the main site of the World Trade Center, that it would remain commercial. But having residential tenants is an attraction for the commercial tenants. They can attract and retain their employees, give them a walk-to-work opportunity. I think that's really important for the future of downtown.

AUDIENCE: I'm a student of Dowling College of New York. It's interesting how I keep hearing that this is a public process, and that it will be based upon public input, but yet, as a person who has attended most of the public hearings regarding the World Trade Center site, I hardly heard anyone speak against rebuilding the Twin Towers. It feels like that has been what the public wanted the whole time, and yet this is not considered an option. And to be honest, the picking of Daniel Liebeskind's plan was not even the LMDC's choice, it was actually Pataki's choice. . .

KATHRYN WYLDE: I think we've got the question: Why didn't we rebuild the Twin Towers? I think for those of us that used to walk in the wind tunnel around the Twin Towers and almost get blown over I think I can give an answer, but Stefan, you were prescient in bringing this up earlier.

STEFAN PRYOR: I think this is part of the point in terms of why we are where we are in our process, and I think it's healthy. The fact is, there have always been points of view at both edges of the spectrum, and different points of view within, as to what we should be doing, and I think that the short answer as to why we are proceeding with the Liebeskind Master Plan is that it was forged through this process, where hundreds and thousands and tens of thousands of inputs were taken at different forums. We've held a couple of hundred public sessions, including probably half a dozen just in this auditorium alone, where people expressed all kinds of points of view. The fact of the matter is that what this has resulted in is a design that doesn't respond to every single point of view, but incorporates what many people wanted. Simply rebuilding the Twin Towers would never have been feasible: The imperative to rebuild the street grid would never have been accommodated with the rebuilding of the Twin Towers; the imperative of building a memorial site that's abundant within the sixteen acre site and preserves the site of the original footprints, would have been diametrically opposed with a straightforward move, such as rebuilding the towers. The fact of the matter is that we are five years beyond 9/11, and I think the greatest point of public consensus now is not to revisit fundamental decisions, but to rebuild, and that is what's happening. And that's why John and I and others are concerned that the main issue now is going to be construction coordination. That, too, is well underway.

KATHRYN WYLDE: This must be the end of our discussion. Thank you to Dr. Caputo and Pace University, and to Meghan French, and to our panelists. Thank you.

PANEL SEVEN
HIGHER EDUCATION

JOHN MERROW: My name is John Merrow. We are going to talk a little bit, five years later, about the effect of 9/11 on colleges, and on professors and students. This is going to be a terrific panel, no speeches—just conversation. Let me start by introducing our panelists. Sadie Bragg is the Provost of the Borough of Manhattan Community College. In your program it may say Antonio Perez, President, but we decided we wanted the A-team . . . no, don't tell anyone I said that. David Warren is the president of the National Association of Independent Colleges and Universities. Robert Hackman is a graduate student here at Pace studying Internet Technology. You may want to ask him some of your internet questions this afternoon, but now he's going to talk what it is like to be a graduate student, particularly a graduate student from another country (in his case the Republic of Ghana), after 9/11. And President of Pace University, David Caputo, whom you may have met along the way, is the man I refer to as my boss. What I'd like is to set this discussion in context. David Warren and his group, the National Association of Independent Colleges and Universities, have conducted a survey, basically asking campuses what's happening. David, tell us a little bit about the study, give us some background.

DAVID WARREN: The study went out to the membership of our association which is a little over 900 colleges and universities and is reflective of all the private research universities, like the Ivies, and the smaller, liberal arts colleges.

JOHN MERROW: You're talking about four year schools?

DAVID WARREN: Four year schools, in the main. It's an enormously diverse group.

155

JOHN MERROW: And how many responded?

DAVID WARREN: One hundred thirty-eight responded, a little under 15%. We want to be clear that this is not a statistically significant return. I think it's a highly suggestive one. I'll summarize in a moment what some of our findings are, and I think it will be helpful in suggesting how you want to go forward. A copy of the questionnaire will be on the table after the panel and you'll see the ten questions and the response rate by institutions.

JOHN MERROW: What's interesting, David, are some of the comments. What's the one finding that jumps out at you?

DAVID WARREN: I'm going to fudge and give you two. The first that strikes me as pretty dramatic, and it will surprise people here, is the effect of September 11 on the capacities for international students and international faculty to come to this country, on the one hand, and on the other, the extent to which it has affected, pretty substantially, a number of American institutions who want to send students overseas. We asked people to respond, on a scale, to these questions about the impact of 9/11 on their institution with regard to international students: Was the effect transformative? Was it major? Was it moderate? Was it hardly noticeable at all? Or they could respond, "I don't know the answer." Seventy-two percent of the institutions said that there was at least a moderate effect on international students coming. Thirty-four percent said it was major or transformative, that the visa and so-called SEVIS system, which is a tracking system, has made it much more complicated. Sixty-six percent said the same about faculty, coming from other countries, and a full third of them, thirty percent said the transformative effect was major. If you look at the data, you will see from 2001 to date the downward trend of international students coming to this country, as well as faculty. There's a small irony here, which is that the number of American students going abroad has actually gone up even though half of the institutions said they're having a problem. Now, almost 200,000 are going abroad.

JOHN MERROW: Tell me a little bit about that. What do you attribute that to? Why would more American students want to go study abroad?

DAVID WARREN: When we did a review of the literature to design this, we tried to pull out some themes. One of the themes that came through was a sense of global orientation that students did not feel was part of their education. Many believed that September 11 was so clearly a signal that they had to think in a new and global way. So colleges have begun to say, let's see how it is that it would be possible. Robert Hackman's undergraduate work was at Goucher; it's the first institution in the country that now requires every student to spend a year abroad.

JOHN MERROW: A year abroad used to be, well, you go to France. Are you saying students are going to Muslim countries?

DAVID WARREN: Students *are* going to Muslim countries. Of course, English speaking countries are the line of least resistance, so Australia and New Zealand and, of course, Great Britain are popular. But what we see in the growing number of students going abroad is that they want to be engaged in Middle East-related questions; they want to be oriented with Muslims, oriented in Islamic issues.

JOHN MERROW: David, have you seen that latter effect here, students at Pace saying we'd like to get out of this country, get out of our chauvinistic, if you will, or nationalist enclave?

DAVID CAPUTO: We've had a significant increase in the number of our students who are going abroad for a variety of experiences and we think it is because of our emphasis and our strategic plan on internationalization.

JOHN MERROW: Is this post 9/11?

DAVID CAPUTO: This is post 9/11; it is not directly related to 9/11 but more part of a strategic plan. I would like to go back to a point on the immigration issues. I think the subject of visas—and it is very problematic—is but the tip of an iceberg of something that is far more serious, and that is that the United States faces the very real possibility that it is going to lose its position as a pre-eminent leader in higher education.

JOHN MERROW: Why? In what way?

DAVID CAPUTO: It's making it more difficult for us to attract students and at the same time the competition is much greater around the world, from all sorts of other institutions both in Great Britain and Australia, but also, I think, there are more and more institutions that have developed a much more competitive model.

JOHN MERROW: Chapter and verse, you can say we're losing these graduate students.

DAVID CAPUTO: Not just graduate students, undergraduates as well. I always argue that the undergraduates are every bit as important as graduates. We lost approximately 40-45% of our international students.

JOHN MERROW: How many students?

DAVID CAPUTO: We went from about twelve hundred to seven hundred.

JOHN MERROW: So it hurts you in the pocketbook?

DAVID CAPUTO: It's hurt in the pocketbook, but far more important, I think, is that it's hurt us in terms of the richness of the diversity that we have on campus—not just in terms of religious and ethnicity, but the intellectual and academic diversity also is missed.

158

JOHN MERROW: One thing that David's study suggested is that it also created a bureaucracy, if I read it correctly. A number of campuses reported back saying, "Well, we've had all these different forms we have to fill out so we have to add staff." Is that correct?

DAVID CAPUTO: Yes, we've had to deal with that in terms of trying to provide greater support and assistance. This is a very complex process to start with, the new rules and regulations make it even more complex. International students, and especially new international students, need a great deal of time and attention.

JOHN MERROW: Robert, you came here with four years of high school and took four years of college. Now you're a graduate student at Pace. What's the process to renew your visa?

ROBERT HACKMAN: You have to go home, you have to get out of the country, you then have to go to the U.S. Embassy and apply and there is no guarantee. I was very scared that I wasn't going to be able to come back into this country because my visa expired after my freshman year of undergrad. I wasn't sure I was going to be able to come back because of all the new policies and what was going on. That is something that every international student has to deal with: if I leave this country, I might not be able to come back.

JOHN MERROW: So did you sit around for weeks waiting?

ROBERT HACKMAN: I wish I knew exactly how the process works. I feel that's part of the problem. There's a whole bunch of new regulations, new policies that we had no idea about; we were kept in the dark about a lot of things. Even our international student advisor at our school, our undergraduate advisor, wasn't really prepared for the changes.

JOHN MERROW: Were your professors in the dark as well, with the new rules coming out? How did that work?

SADIE BRAGG: I think that at the Borough of Manhattan Community College, we had 18,000 plus students and roughly 2300 international students, probably the sixth largest community college in the country in terms of international students, which means that we do not actively recruit. But since 9/11, with SEVIS and the other kinds of processes that have been put in place for students to study in America, all that's changed because of paperwork. And yes, we had to add staff, to our office of admissions, to deal with international students.

JOHN MERROW: Which is what the study says.

SADIE BRAGG: Yes.

JOHN MERROW: Robert has suggested that you're sort of in the dark, from the administrative level. Do you find that with all groups?

DAVID CAPUTO: While the rules were being developed, over a process of probably fifteen to eighteen months, I think everyone was uncertain. You just had to go along. I would like to add that this is a problem not just for our international students, but also for the international faculty.

JOHN MERROW: Is it tougher for international faculty than students?

DAVID WARREN: From the institutional perspective, the study suggested that it was slightly less difficult for the faculty—that is, 30% of the institutions said it was a great or transformative problem.

JOHN MERROW: I'm wondering if there's a differentiation in faculty—easier, perhaps, for someone teaching about literacy, but less easy for a political scientist?

DAVID WARREN: I only have some of the more anecdotal observations. One of the problems in the system that was referred to by Sadie, is SEVIS, which is a tracking system. Any time an international student enrolls, they have to tell the institution when they arrive, what courses they're taking, where they are living, where they move about in the country, what their progress is on their degree. It's exceptionally time consuming. This does raise some questions of whether it's being too intrusive, and it does create a database that has problems.

JOHN MERROW: Okay, let me play off intrusive. I'd like to know from Robert whether knowing that you have to leave the country and reapply, what, if any, impact does that have on your own behavior while you're here. For example, are you reluctant to be political?

ROBERT HACKMAN: If I were from a Muslim country, I would be. Even I don't really like to talk about how I feel about America, because it almost seems that if you have anything contrary to say about the government's policies, it's not a welcome view. That is, you're not supposed to say anything, that is negative in any way, about America's policies.

JOHN MERROW: You say you're not supposed to. Is that something you would intuit? Or is it kind of in the air?

ROBERT HACKMAN: Kind of in the air . . . you just feel like you're not supposed to.

JOHN MERROW: This has to be upsetting for these two college presidents.

SADIE BRAGG: There's another issue for students who have actually been in America and are not going back to their country, those who live here and are going to school. Since 9/11, it's been extremely difficult for those students to voice their opinion about what they think politically, particularly those who are Muslims, and those who wear their Muslim garb. In fact, after

9/11 we lost a significant number of Muslim students because the campus is right in the shadow of the World Trade Center. It was very difficult for them to even explain to us as to why they were leaving. But it was difficult to get on the train, to come from Brooklyn and other places.

JOHN MERROW: So, a physical fear.

SADIE BRAGG: Yes.

JOHN MERROW: But I was talking about ideas.

SADIE BRAGG: There is, of course, also an idea fear. They won't speak when they are in class, they will not engage in certain dialogue. Even among our faculty it is difficult. We hold small group discussions with the faculty and the president. Our most recent discussion was on globalization. There was a very diverse group of individuals in the room, yet there were certain conversations where they were careful about what they said, because one faculty member was from Pakistan and another was of Jewish descent. So the academicians are as equally cautious as students about what they say.

JOHN MERROW: David, it must bother you.

DAVID CAPUTO: Certainly it does because the academic institution should be a crucible where ideas, regardless of what type of ideas, have a chance to be freely expressed and openly discussed. I do think there has been some dampening of that over the last couple of years. I'm not sure how much of it, though, is something that is intuited on the part of the individual rather than in terms of stated policy.

JOHN MERROW: Self censorship versus . . . in the end it's the same thing, isn't it?

DAVID CAPUTO: Well, it depends. I think self-censorship may last less long and may not be as broad in terms of the topic covered. I do think that the difficulty is trying to get the sense of dialogue and interaction that you want started on any of the major issues that we face. We find in such a divisive time that quite often most student discussions degenerate into very similar things that you see going on on the national level, where people are not able to have a sensible discussion. That's one of the things that I think all of us pride ourselves on, and all of us do everything we can to make sure that that continues.

JOHN MERROW: The campus should not be afraid of any idea. Let me turn to David Warren. One of your questions was about the impact of 9/11 on academic freedom, wasn't it?

DAVID WARREN: It is, to borrow the phrase "the dog that didn't bark," the most surprising finding. By that I mean that on the scale that we gave them, eighty-two percent of the institutions responding said there was very little or no effect on academic freedom. Now, that doesn't strike me as the experience that I've had. It seems to me that there has been a diminishing feel for a free, forthright, and frank exchange. If you link that to the question we asked about the Patriot Act, it's a little mysterious because sixty-seven percent said there was little or no effect generated by the Patriot Act.

JOHN MERROW: So you're saying you don't believe those eighty percent?

DAVID WARREN: I find it curious, at best; it just doesn't square to my experience.

JOHN MERROW: We've got plenty of examples—Churchill in Colorado; a guy in New Mexico; a case in Georgia; a case in Florida—where people have made outrageous statements.

DAVID WARREN: And, again, at the top end of the scale, only twenty-seven percent said that there was a great or transformative effect or consequence of the Patriot Act. So, among other things, what will happen now is that some version of this questionnaire will be taken out into the field where we can test it. This was, by the way, done online, by email sent to all the presidents. If you go to the link or to our website you can answer the ten questions. We wanted it to be impressionistic: don't take more than ten minutes, don't do any research, you're the president, just give us your take. And that is what we got.

JOHN MERROW: We've got two more questions, let's do the same thing. Effect on academic freedom?

SADIE BRAGG: Well, as I said earlier about this, I don't think that academic freedom, as we define it in academia, is a problem. At least in our college, from my experience, it wasn't. People have an opportunity to speak out as they always do. I do think that they were far more reserved about their conversations. I'm not sure that that was curtailing of the academic freedom to say whatever you wanted or whether it was a level of respect for colleagues in the room. When there were different ethnicities, you could see that there was a limitation, somewhat, on the freedom of what was said.

DAVID WARREN: We've not seen an increase in complaints from either students or faculty in terms of their having their free speech or academic freedom infringed upon. I think what we have seen is, especially on Arab-Israeli issues, a heightened tension, which could go back to 9/11 but probably goes back much further.

JOHN MERROW: It seems to me that that is a different issue. Let me pose it this way to all four of you. I was talking last week to a high school student and he wanted his class to ask the question, "Why is the United States hated abroad?" But, he said he was scared to, and did I have any advice on how we could ask that question. Now, I don't know if you remember that after 9/11 there was a famous picture from some Muslim country of protesters holding up a sign that said, "America, ask yourself why so much of the world hates you." And it struck me that there wasn't a school, high school, or college in the country where they could actually do that, without fear of losing their jobs.

DAVID CAPUTO: I'd be very surprised if that were the case, if there were college campuses where that question could not be raised, especially within the context of an appropriate class together.

JOHN MERROW: Do you feel the same way?

SADIE BRAGG: I think it can be raised, but I believe that people have reservations. I think it definitely will be raised because in academia people feel free to say what they want to say, but I do think there would be some reservations.

JOHN MERROW: And would you hear about it? If I worked for you, on your faculty, and I gave my class an assignment to write something about this question, would you hear about it?

SADIE BRAGG: I would only hear about it if, in fact, the faculty person that was leading the discussion took a side on it. You cannot, as faculty, decide on the conversation. It has to remain open; you cannot state what you think. The moment you enter the conversation, then it becomes a different conversation. Because you're the one who's giving the grades to the students and all of a sudden you are saying that you think the students need to mean the same thing as you. I think it is quite healthy for faculty to start this dialogue, but they should remain neutral in the conversation.

JOHN MERROW: Do you think that you should engage in this topic?

SADIE BRAGG: Oh yeah, I think that it should happen in the academy. Where else can you have this kind of conversation? It definitely should happen.

JOHN MERROW: Well, it should happen, but David Warren says that though people say it is, he doesn't believe it.

SADIE BRAGG: Well, I was surprised by the data. When I read this report, it was the one thing that didn't seem quite right to me, either. Not that it was overwhelmingly the other way in my college, I just would not have expected this.

DAVID WARREN: And John, there's a clue as to why I think that number is soft. We asked the question, "What's the effect of the Patriot Act?" Now, twenty-seven percent of the institutions said it has been great or transformative.

JOHN MERROW: But it had no effect?

DAVID WARREN: Apparently there's been no effect on academic freedom. Well, we know that the FBI can now surveil folks, look through the archives, tap phones, and be engaged in surveillance of the internet, all under the rubric of the Patriot Act and consequence of Sept. 11, so holding these two ideas together suggests a little cognitive dissonance. I'm betting that it had some spillover greater than the survey suggests.

JOHN MERROW: Robert, you're a technology major, do you feel the post-9/11 effects on books you take out of the library?

ROBERT HACKMAN: Well, I was thinking of an incident where an Egyptian student was telling me his thoughts on the whole Arab-Israeli conflict and he was giving me some websites to check out, and I felt a little hesitant to go check out those websites because I wondered, in the back of my mind, what if this is being monitored?

JOHN MERROW: Have you talked about that with other graduate students from outside the United States?

ROBERT HACKMAN: Not too many.

JOHN MERROW: So you don't know whether your sense is widely shared?

ROBERT HACKMAN: No, I don't.

JOHN MERROW: Why wouldn't you talk to them?

ROBERT HACKMAN: My Egyptian friend is very passionate about this issue, and his views are totally different from what the majority view is.

JOHN MERROW: I lived through Vietnam, and there were tons of protests. Now, in terms of 9/11, and the war in Iraq, it's non-existent, can somebody connect the dots for me why? Is it just that we don't have a draft, and your students don't worry about getting blown away? Or is there some recognition that this war is different, because it's called a war on terror? Can somebody work that through for me?

DAVID CAPUTO: I'll take a first stab at it, and I'm sure my colleagues can add to it. I think the contextual setting is very different. I think the anti-Vietnam War effort grew out of a total sense of frustration of governmental policies in general. It was a sort of culmination or mid-point of a major effort to redefine American society by large numbers of individuals. Whereas, many students today are very concerned about their economic well-being and the future. I think they also lack a sense of history.

JOHN MERROW: Is it the draft, or rather the lack of a draft?

DAVID CAPUTO: I think the lack of a draft for many is obviously a factor; it is not necessarily in their self-interest to argue against a policy that in fact doesn't draft them.

JOHN MERROW: It's remarkable, really, given the profound effect the Iraq War has on the life of so many people, with the

National Guard and the Marines now being involuntarily called up for a fourth tour.

DAVID CAPUTO: It may also be early in the process, four years. It's younger than Vietnam.

JOHN MERROW: It's longer than World War II.

DAVID CAPUTO: But you haven't had a sustained escalation as you did in Vietnam.

SADIE BRAGG: I also think of being a child of the 1960s, and having been on a college campus in the 1960s, that there were so many other issues going on. And Vietnam didn't come to America, we went there. If you think of it that way, like those of us who watched the plane go through the World Trade Center . . . you have to stop and think. I don't think people are as angry as they were in the sixties.

JOHN MERROW: Even if there were a draft?

SADIE BRAGG: I agree, it would then be a different situation.

JOHN MERROW: Are you surprised that there's virtually no political protest about this war in Iraq?

ROBERT HACKMAN: Among the students?

JOHN MERROW: Yes. Is that a subject of conversation?

ROBERT HACKMAN: Well I went to Goucher College for undergrad. It's a liberal arts college. It's the sort of place where protests would take place. But I felt almost a sense of being desensitized to the issues. Of the issues being too large.

JOHN MERROW: Interesting.

DAVID WARREN: I was subject to the draft in the 1960s and I was drafted. I was ultimately excepted for one of these nonsensical reasons—-different part of the United States Government wanted me to do something else, go on a Fulbright, so I was preserved. Conflicts on campus were so multitudinous—race, gender, the whole concept of whether institutions reflected integrity—democracy was coming apart. These conflicts were all going on, the draft was there. But there was, also, a really different phenomenon going on—you could be involved in all this, but you could get a job the next day, if you wanted to. Economic determinism was not at work; you were pretty certain you could find a job if you wanted. But the last difference is context. On September 11, 2001, you had a circumstance where you saw a plane go into the building. People in *this* auditorium were caught up in this. My daughter lives in New York. I told David yesterday, if her alarm had gone off, like it was supposed to, she would have died. But it didn't, so she didn't get up in time to go the office, which was destroyed. That has a visceral, psychological, and political effect on you that is pretty transformative versus the circumstances of which I have such clear recollections, in the 1960s.

DAVID CAPUTO: I'd like to add another perspective. I think, in many respects, today's student is channeling a lot of that energy into their own volunteer efforts. I mean all of us are finding, on all of our campuses, that students are much more engaged with their community, with community groups. This is not always easy, but it's certainly easier than it was in the past to get students mobilized to help. I think of the student response to Katrina, in terms of the large number of students who spent vast amounts of time there helping very diverse communities. It is very laudable, and I think you can cite that in a number of cities.

JOHN MERROW: We're talking about the effect of 9/11 on campus. Is there any sense among any of you, of a sort of fear of "Big Brother?"

SADIE BRAGG: I won't say to you, "don't worry." I think that those of us who are in charge of various things on campus, including presidents and student affairs people, have to protect our students. Yes, you do worry about the amount of data you generate on some students that could be harmful to them in the hands of others. If you've never worked with SEVIS, you should know that it is a repository of a lot of powerful information. In the hands of the wrong people, it could be interpreted in many ways. You have to establish rules and regulations.

In some ways, I don't know if I'd use the term "worry", but I would say that I'm extremely concerned about the information and how it's used. I do want to protect our students.

DAVID CAPUTO: I would say I share your level of concern, but not just because of a future attack. I think one of the major issues related to the increased technology is an erosion of all of personal privacy, and the fact that that information can be used in a variety of ways.

JOHN MERROW: But that's not related to 9/11, that's just . . . progress.

DAVID CAPUTO: But it is related to 9/11, because it's technology that was developed since, and has been developed partly in response to, 9/11. So I think it's very difficult to separate the two. I think 9/11, especially for those of us in the city, had some very specific short-term impacts and very still diffused long-term impacts. But when you think about it, the class of 2010 were eighth and ninth graders at the time of the attack, and we know what happens when you ask students today about the Vietnam War. It is the equivalent of going back almost to the First World War. There is a very short time-span in a lot of psyches. I think there is concern about the technological capabilities now, whether it be the camera on the corner monitoring your behavior, or the traffic camera that's taking photos, as you, for whatever reason, run a red light. I think there is increasing concern about what you called Big Brother, and I think most people are concerned about the governmental activity to raise it, and I think it's been heightened by 9/11.

JOHN MERROW: That reference Big Brother may not mean anything to a lot of people here— it's from an old novel, *Nineteen Eighty-Four*, by George Orwell. He wrote it in 1948. I do think Big Brother is watching you. It's a scary notion. Robert, you're a technology guy, do you worry? Do you worry about this idea that somebody's watching you, that the government is watching you?

ROBERT HACKMAN: Well, when the whole issue came up about wiretapping the phones, it was only because the information got leaked that the American public knew that their phones were being tapped. And so I feel, even for American citizens, that it's becoming an issue, and definitely more so for international students.

JOHN MERROW: But some Americans said there's nothing wrong with that. The only people who should be worried are people who are saying bad things.

ROBERT HACKMAN: I feel, for most students, that there is definitely a heightened sense that, "my liberties are being taken away, my privacy is being taken away."

JOHN MERROW: You're talking about foreign students?

ROBERT HACKMAN: Both international and national.

DAVID WARREN: Some of you may have heard the really compelling presentation by Lee Hamilton earlier. He said that our Congress today has not maintained the checks and balances. And that, yes, there are legitimate security concerns, but on the other side, there are legitimate civil liberties concerns, and right now, they're out of balance. We need to restrike that balance. I'm pretty clearly on record as opposing something that's being recommended by the Secretary of Education called "student unit record data." What that means, very simply, is that all sev-

171

enteen million students in the U.S. would go into a database. If you apply for federal financial aid, all of your parents' financial data will be thrown into that database; it's going to reach back to K through 12, and it's going to go forward to follow you in your employment. Now the purpose of this is to be able to have sufficient data to answer questions about graduation rates and about transfer of credit. I think it raises substantial problems of privacy, of confidentiality, and of the possibly inappropriate use of that database for other purposes. It's going to be a very significant discussion in this Congress as to whether or not to provide the funds to enable this to take place.

JOHN MERROW: So this is something the Bush Administration wants to do?

DAVID WARREN: To the extent that the Secretary of Education is an exponent of the Administration, absolutely. It is, by the way, a very interesting issue because it has split the Congress. John Boehner, who's the Chairman of the Committee of Jurisdiction for higher education, and is now the majority leader, along with a gentleman named Buck McKeon both oppose this on grounds that it is a violation of security and confidentiality and that it will be used for inappropriate purposes. What does it mean to throw into a database all of this information about an individual and to what ends will it be put?

JOHN MERROW: You sound a little more confident that the Congress will step up to the plate and at least debate it.

DAVID WARREN: I trust that is so, but the House of Representatives may change hands, and we'll find out what the consequences are, if the two most outspoken opponents of it move to the minority. I haven't placed my bet yet on this one.

JOHN MERROW: David, there's an eleventh question in your survey, an offer to expand on any of the first ten questions. I was struck, in particular, by one comment, and I'd like to ask all of you to react to it. We don't know which president said it, but he or she wrote: "The significant change has been in the attitude of the students, who are more openly patriotic, more inclined to think critically of terrorism-related issues, and much more likely to support and admire those who served in the military." Does that surprise you?

DAVID WARREN: Not if you look at one question that I didn't talk to you about, which was about the effect of 9/11 on the curriculum. Fifty-eight percent said moderate to significant effect has taken place. Twenty percent said it was major to transformative. What does that mean? I think students responded constructively and positively to the events of September 11. They want to know about the Islamic faith, they want to know and learn more about the Middle East. They want to travel abroad; they want to take comparative religion courses. When you get that level of change in the curriculum in American higher education, that's substantial.

JOHN MERROW: Wait a minute, doesn't that fly in the face of what the other three people either said or implied that, less willing to talk about controversial issues, pulled back?

DAVID WARREN: Not necessarily. I don't think they're mutually exclusive at all. I think students are very interested in inquiry and talking about these issues, on their terms. It's entirely fair to raise questions about what is happening in the Middle East without having to assert a particular political proposition. But I thought that was the most promising piece of data in the entire survey, that the curriculum has moved, that students are much more inclined to volunteer, to go overseas, and to think more globally.

JOHN MERROW: That is heartening.

SADIE BRAGG: I totally agree with David. Globalization has actually transformed our campus. Our students and our faculty are excited about embracing all kinds of ideologies, all kinds of countries and beliefs, like they've never been before. So I don't want my previous comment, about what faculty would say in class with a student dialogue, to be a final statement. That was a different conversation. Students, now, are more intrigued than ever when they hear about other religions, about how things happened in the Middle East. We have a very active study abroad program. We've sent thirty students, fifteen for the last two years, to the Salzburg Seminar in Austria, and it's been phenomenal. They've come back extremely transformed about what they believe and what they thought they understood, but really didn't understand, as Americans. So I honestly think the curriculum *is* changing. As a person who's directly responsible for a lot of that—a lot of new courses, a lot of new ideas—I think it will allow the next generation of students to be a lot more embracing of a variety of ideas and ideologies. And maybe they won't quite feel the way Robert feels about what he can say and not say, because they'll all begin to understand each other. I think if we had more dialogue like that on our campus, students would not be afraid to go back.

DAVID CAPUTO: I would like to respond to your comment about the increased respect for those who serve in the military. I think that there *is* a heightened respect, especially since most students realize that it's a voluntary decision. And I can tell you, we have lost students. I'm sure that's true for campuses across this country. We have former students, or students who left in order to enlist and serve, who have subsequently gone to Iraq or Afghanistan. There is a palpable feeling of loss across campus that I don't think we felt before. And I think part of it is that our sense of loss has been heightened since 9/11 in general. I think 9/11 is very personal to everyone, to all of our students and faculty. I think we all feel that we're here as educators to build lives and to open doors. And war, whether it be an act of terrorism or a military war, closes doors and ends lives.

JOHN MERROW: So you're sensing your students are really pulled in to what's going on?

174

DAVID CAPUTO: I think our students have a greater sense of understanding and concern about what is going on around the world. If you gave them a set of questions to answer, I'm not sure that they necessarily would be able to score better, but I think that they would have a better sense of unease about the times that they are in.

JOHN MERROW: Now an awful lot of people expected there'd be a big jump in the student vote in the last presidential election. There was a big "get-out-the-vote" effort, but it didn't happen.

DAVID WARREN: Well, yes and no. We saw, from 2002 to 2004, an eleven percent increase in voting on campus.

If you're talking about eighteen to twenty-four year olds on campus, they turned out to be the single highest registered subset in the country, and of that group, about seventy percent vote. So it's a pretty substantial number of persons who did vote in 2004, greater than the elderly.

JOHN MERROW: So you're saying they vote?

DAVID WARREN: They vote about seventy percent, and they register at eighty-eight percent. It's a very substantial reflection of the group.

JOHN MERROW: So what you're saying is that they're politically aware, but they're not out carrying banners.

DAVID WARREN: Volunteerism is the way.

DAVID CAPUTO: We all think they've invested a lot of their time and effort into their community and community groups. I think the concept of public service takes precedence, and I think it's partly due to both their lack of knowledge about the political process, but also due to their concern and perhaps suspicion of political leadership right now. And I think groups such as Project Pericles, Campus Compact, and others, are really trying to engage students, and they've been very successful. Each year,

175

when we've done our voter registration drives, we've seen a modest increase in the number of students registering. Anywhere from eight hundred to twelve hundred students were registering. Sometimes it's a little higher or a little lower. We have no way of tracking whether or not those students actually vote. But we know we're able to get them registered.

JOHN MERROW: We only have a couple of minutes left. What do you worry most about, in the context of 9/11 and its impact on your lives? What do you worry most about?

SADIE BRAGG: There are several things. When the World Trade Center was hit, we didn't have a lot of plans in place to be able to deal with the tragedy, now we're prepared. Just the day before coming here, I went to a CERT ceremony where we deal with community emergency response training. I think that what we concern ourselves with now on our campuses is whether or not we are prepared.

JOHN MERROW: For the next attack?

SADIE BRAGG: If and when there may be another attack. There are many things that we've learned, first-hand, from having lost a building, from having been out of our building for three long weeks. From being taken over as an emergency place, from having a gym, literally being turned into a morgue. I mean, there are many, many things that we've already experienced. I think that, as a college, we are preparing ourselves so that we won't be in the position that we were in.

JOHN MERROW: My wife is the head of a school where we live in California, and she has stocked three days of supplies. It used to be a school principal had to think about how the kids would get out. Not anymore. Now you have to say, "What if we have to keep them?" So they've stocked away beds and blankets and supplies.

SADIE BRAGG: It's just like during 9/11. The most important thing that happened to us was that the engineer who was in the building before the towers fell was able to predict what was going to happen, literally, and shut off all of the outside air coming into the building. Everything was totally shut down. Had we not done that, considering where we were located, we would have been out of our building much, much longer.

JOHN MERROW: So when I ask you what you worry about, you worry about that sort of nitty-gritty detail?

SADIE BRAGG: Very detailed things. As a person who's working closely with the president and with our board administration, I worry not just how to get our students out of our building—I mean we can do that pretty well, but how to save our building so we can come back to it.

JOHN MERROW: David?

DAVID CAPUTO: Being located across from City Hall and the Brooklyn Bridge, and around the corner from the World Trade Center site, we have security concerns that we are constantly reminded about. And those are important. We *have* spent more on security; we've had more discussions about security than in the past. But to me, the thing that I worry the most about, is that the discussions and the policies related to the War on Terror and so-forth have so sharply divided the country. The country is so divided right now that I'm very concerned that we've lost sight of what I think is the single most important issue: education. Education, from pre-K through post-graduate. We're simply not investing, as a country and as a society, in education as we need to be. That will do far, far more to undermine us as a nation and as a society than any terrorist attack would.

JOHN MERROW: Fragmented, divided, maybe polarized.

DAVID CAPUTO: I fear we are losing sight of what has always been an essential aspect of the American dream, education.

177

JOHN MERROW: Okay, I'm going to take it now to my graduate student from another country, but studying now to get a masters' degree here. In the context of 9/11, what do you worry about most?

ROBERT HACKMAN: Well, it's been very stressful for me just having to deal with all the laws and all the changes. One of my concerns is that things are just going to get more stressful. There are going to be more regulations, more policies, and it's going to be very difficult, even more so than now, to be an international student in this country.

JOHN MERROW: Several people have said that there's a new attitude that many Americans want to know more about the rest of the world. Have you seen that?

ROBERT HACKMAN: I would agree with that. To comment about the opening patriotic part though, I would say that that would depend on where you are. Because I have friends, American friends, that say that when they went abroad, on study abroad, they were afraid to say they were American.

DAVID WARREN: One of my three children lives in Barcelona, and I have brother living in Munich and a sister living in Brussels, and so I hear from them this perception that they love Americans but feel America is somehow off-track. David, you're a college president, I'm the head of a large association of independent colleges and universities. What keeps you up at night in the context of 9/11?

DAVID CAPUTO: Let me echo David's observation. A partnership came together, in 1965, when they created the Higher Education Act. It was a partnership of the federal and the state governments, foundations, corporations, the American family, and colleges whose goal was to make college possible for needy kids who were bright. That partnership is coming unraveled.

JOHN MERROW: But that's not 9/11.

DAVID WARREN: Well it's very much associated with 9/11. The Pell Grant, which is key for our neediest kids, has been flat for five years. It went up seventy-four percent between 1995 and 2000. Since then, it has not moved a penny. If we had the funds that had been given over to Iraq and terrorism we could have doubled the Pell Grant in a month and a half. We've been trying to double the Pell Grant for twenty-five years.

JOHN MERROW: You're saying a month and a half of spending in Iraq. . .

DAVID WARREN: . . .would double the Pell Grant. And so we're looking here, my friends, at priorities and the redistribution of resources, and the war is having a profound effect on the capacity of folks to go to college.

JOHN MERROW: The Pell Grant initially paid about ninety-five percent of the tuition of state college or university. Today it's not even fifty percent.

DAVID CAPUTO: It's less than fifty percent.

JOHN MERROW: Which means kids are getting priced out of college. So you're saying that your worry, then, is that post-9/11 resources have been focused elsewhere.

DAVID WARREN: It's exacerbated what has been a trend.

JOHN MERROW: But nobody said their big worry is the government being so intrusive. That's not the big thing, but it's there. Do you see an elephant in the room?

DAVID WARREN: Well I've got a herd of elephants in my room, and that's one of them. Funds for higher education would be another one. I'm perpetually concerned about regaining a balance between security and civil liberties. We've got to get that

179

right. It was the most hopeful thing I've heard, that somebody as respected as Lee Hamilton was in agreement.

JOHN MERROW: I want to thank all four of you for your thoughtful comments, and I want to also express my appreciation to the audience for being here.

PANEL EIGHT

AMERICA'S PLACE IN THE WORLD

BEVERLY KAHN: Good morning, and welcome to what promises to be a very important panel on the topic of America's place in the world today. My name is Beverly Kahn, and I am Vice President for International Opportunities here at Pace University. I am a political scientist by training and former Fulbright Scholar. I specialize in European politics and political philosophy, including democratic philosophy. In my thirty-three years as a professor and administrator, I have worked at the University of South Carolina, Ohio State University, Fairfield University, and now Pace University.

I arrived at Pace two weeks before 9/11. It is in emergencies, in times of crisis, that individuals and institutions show their true colors. I am proud of Pace University. I was a newcomer on the scene and I saw how Pace responded with rapid and smart decision-making and with moral leadership.

The freshmen of 9/11 graduated last year. Those qualities that Pace showed in the emergency after 9/11 are qualities that we, in the strongest and best sense of the liberal arts tradition, attempt to nurture in our students here at Pace. We are celebrating our Centennial, our 100 years. But we also are looking ahead to our second 100 years and internationalization is an important component of our strategic agenda for our second century.

We want to prepare our students to be global citizens. We are working diligently to help them understand the interconnectedness and interdependence of the world of nations. In the last three years, we have increased the number of students that study abroad by 192 percent. Now, seventeen percent of all Pace undergraduates, during their career here, study abroad or have another significant international experience. We teach eight modern languages, including Arabic, Russian, Chinese, Japanese, and Portuguese, as well as Spanish, French, and

Italian. And we want to teach more. We also want to welcome international students here.

We have 192 students at Pace on non-immigrant visas and another 800 students on immigrant visas who have refugee status or are green card holders. We have students from 110 nations. Our faculty are eager to be involved in international research and collaboration. We are working particularly hard to make connections with China, Brazil, and Italy, as well as with other nations around the world. That is part of what education is about. I'm proud of my institution and what we are doing under the leadership of President David Caputo.

We are here, today, to talk about the world and America. And we have some distinguished panelists with us. I would like each now to introduce themself and then we will begin with opening remarks by Professor Kenneth Jackson. Kenneth, why don't you introduce yourself for us.

KEN JACKSON: I am Ken Jackson. You will pick up quickly that I am not originally from New York, but from the South. I came here 38 years ago to teach Urban History at Columbia University and I've been teaching and writing in New York ever since. One of my books is called *The Encyclopedia of New York City*; it's probably the largest one volume book ever produced [on New York City]. Another is entitled *New York Through the Centuries*. On September 11, 2001, I was the President and Chief Executive Officer of the New York Historical Society and teaching at Columbia. I put that institution on course, at least at the time, to make 9/11 a major focus. We were quickly on the scene collecting all sorts of memorabilia and documents and films and have hosted a number of significant exhibitions and a number of public programs. That is the reason I am here.

BEVERLY KAHN: Thank you, Ken. I would next like to introduce Nikki Stern.

NIKKI STERN: I am a here for a couple of interesting reasons. I'm the former Executive Director of the Families of September 11. I am also a communications professional; I have a consulting

182

practice. I'm also a family member of a victim of 9/11. September 11, 2001, changed so much for so many of us in and around New York City. It's ironic to me that according to the *New York Times*, life has returned somewhat to normal for a lot of people in the United States. I think that it would be false to assume that 9/11 didn't affect, to some extent, America's role in the world or at least the perception of its role within and without its borders.

It seems to me that after 9/11, we had a unique opportunity. I know that's a strange word . . . but we did have a unique position in the world. With the exception of a few ideologically driven pockets, there was an outpouring of sympathy and good will. Five years later, we appear to have reached a point where so many are ready to think the worst of us. Even the good that we do, our traditionally generous approach, seems to have been overshadowed enormously by the impression that others have of our foreign policy, which so many people see as one that favors action over negotiation . . . a sort of "shoot first and ask questions later" approach.

I have heard in my travels from people whom I've talked to that their view of our foreign policy, despite our traditionally well-known largess in the world, is that it is careless of human cost, thoughtless when it comes to local politics, and disinterested in the customs of other peoples or nations. That's very disturbing to me. It's also disturbing to me that even someone as traditionally, if not optimistic at least with a somewhat broad and hopeful outlook, like Tom Friedman can point out in a recent interview that we *used* to be a country that exported hope. Now, we seem to be looked at as a country that exports fear.

What I'm hoping we look at today, and what I want to know from my perspective, is how we got to that position. I was at a conference last year that Steve Clemons's group hosted on Terrorism, Security, and Democracy and was with a group of very esteemed academic presenters and people. I had broached the subject of democracy promotion—something that people were very enthusiastic about and very focused on. I said, "Look, I'm a marketer. Looking at it from a marketing standpoint, are we taking the time that we need to actually understand this target audience and make sure that we know how the product, in this case democracy, would be received? Are we aware that it may be adapted? Or that we may have to adapt the product or accept the adaptation of the product?" My approach was met with some amusement, I think. But of course, a year later, after a number of elections, we've seen that the results aren't always what our policy makers are expecting.

So then the question becomes what is our role in the world? What is it we are supposed to do? What is it we have the capability of doing? Has our role changed, and does it have to change? Those are some of the things that I hope we talk about today. Thank you.

BEVERLY KAHN: Thank you very much. We will try to get to all of your questions and more, as well, because they are important. To my far right is Steve Clemons.

STEVE CLEMONS: I'm Steve Clemons and I head the Foreign Policy program in a think tank called the New America Foundation in Washington, D.C. I'm better known, actually, as a blogger. I write a fairly well-read political blog for political junkies, called *The Washington Note* at *thewashingtonnote.com*. The New American Foundation recently brought in, from the Rockefeller Brothers Fund, a project called "US in the World." It's at *usintheworld.org*. It's a compilation of thoughts and missives prepared by over four hundred intellectuals and academics about how to talk about foreign policy in this post 9/11 environment. It's designed to help people figure out, without bias, how we actually work through and think critically about some of the big challenges.

This is a major time, I think, of discontinuity in the world. What we're likely to do tomorrow is increasingly less related to what we did yesterday. That requires us to rethink a lot of our assumptions, about America's place in the world. It's great to be here.

BEVERLY KAHN: That's wonderful. We have so many terrific public intellectuals here at this conference and we will continue that discussion today. Thank you. Finally, I would like to introduce Alice Greenwald.

ALICE GREENWALD: Good morning. You know, I think I'm the one person on this panel who, when I received the invitation, thought, "why am I being invited to this discussion?" I'm not a foreign policy specialist. I've spent the last thirty years of my life in museums. Directing them, creating them, managing them. Then I thought about the museum that I now have the privilege of helping bring into being: the World Trade Center Memorial Museum.

It's pretty clear to me that the questions that I and my team have to ask, in creating this institution, have to do with what it means to live in a global community at the beginning of the twenty-first century. The Holocaust Museum, where I spent the last nineteen years of my professional life, documents and memorializes a series of events and a catastrophe of humanity that did not occur in the United States. It happened elsewhere. Unlike that institution, the World Trade Center Memorial Museum will occupy, literally, a site of atrocity. And because of the specificity of that site and what happened there, much of the story we tell has to be about remembering those who lost their lives in the horrendous act of murder and terrorism, that took place on September 11, 2001, as well as the individuals who lost their lives in the preceding act of terrorism in February of 1993.

Having said that, however, the team that I've put together, that is now working on creating the museum, has a phrase that we say fairly often, which is that the world is in the DNA of this project. It was after all the World Trade Center that was attacked on the morning of September 11, 2001. The world watched those events unfold in real time, communally, watching the same footage as it happened, together. There is not a human being on the face of the planet that doesn't have a 9/11 story. I don't care whether they were in Antarctica, in Berlin, or in Okalahoma City. They know where they were at that moment they heard the news. So, the matter of what it means to live in an interdependent world is very much at the heart of the program that we have to develop. Citizens of ninety-two nations were murdered at the World Trade Center on that morning. In some respects, we've been cautioned that this may in fact become the world's first global museum. That's an awesome responsibility.

So, having a clear sense of what that might mean and what it means to have an institution where people walking into it have different inter-narratives, is imperative. Right now, if you go down to the World Trade Center site to the fence at Ground Zero, you will see many foreign visitors. There is a multiplicity of languages spoken everyday at that site. Every one of those people comes to that site with their own inter-narrative, their own set of associations. We know that we live in a world where people understand what happened on 9/11 differently. And the museum is going to have to face up to the challenge of accommodating those differences of opinion.

Those are some of the questions that I will be dealing with in creating this museum. So I am fascinated by the conversation we're going to have this morning.

BEVERLY KAHN: Having introduced our four panelists, I'd like to turn to Ken Jackson to give some opening remarks.

KEN JACKSON: I will try to be brief and mostly offer a series of editorials, because the facts are generally well known. September 11, 2001, was the greatest terrorist act in American history. We hope it will always be true that it was the worst terrorist attack in our history. I think also, however, that it was the greatest single fire response in human history.

In World War II, the firebombing of Dresden or the firebombing of Tokyo would have been much bigger human events. What was different, however, about September 11, was that the largest fire department in the world was concentrating its entire focus on this tiny area. There had never been anything quite like that. A three alarm fire in New York is a big deal. Five alarm . . . the city can go years without a five alarm fire. The World Trade Center went to five alarms within minutes, went

186

to double five alarms in fifteen minutes. Then there was what was called a recall, which is when *all* firemen are called to report to work.

They responded immediately. And as they were coming across the Brooklyn Bridge, to confront what was going to be the greatest fire in American history, they knew they were going to lose people that day. It was the greatest tragedy for firefighters, again, in human history: three hundred-forty-three died on a single day. Another fact is that New York and America were the subjects of great sympathy and affection around the planet for days and weeks after that event.

I want to offer a few editorials that may or may not be true, but I'm going to argue them as true and then we can discuss them later.

First, New York went from being the least American of cities on September 10 to the most American of cities on September 12. How often have you heard someone say, "If you come to America get out of New York City, see the country as it really is." I think that was a common view. Yet, when the terrorists sat in a cave somewhere or other, and thought about how to attack America and what place symbolized it more than anywhere it else, they chose the World Trade Center for it represented American aspiration—and the chance to kill more people.

Second, I would say as an editorial, that many people left the city. I know some who moved to the suburbs as a result of fear after September 11. Others left the area altogether. I remember talking to a realtor in Vermont who said, "Boy, there are a lot of people from New York looking for houses." Many New Yorkers did leave. Many companies relocated.

Sometimes, if you look across the river here, it looks like another giant city is emerging from the mist on the other side of the Hudson River in Jersey City and Hoboken. Still, what I think is most important about 9/11 is what has not changed. Think of the predictions. That the skyscraper era would be over, that people would not ride elevators, that they would not enter into subways because of the fear of terrorism. What is important is it didn't happen.

187

We live in a capitalist society. The way you tell what people want to do is by prices. Just think how we wish we were savvy enough in September, 2001, to buy property in New York City. It doesn't matter where. Property has almost doubled in value around the city since September 11. I take that as a vote of confidence in the city. Just look at the prices.

Third, I think more people will be saved by the events of September 11 than those whose lives were lost that day, as a result of policies adopted after that, not all of which I agree with, particularly in terms of ordinary crime.

In London, it's now said that your picture is taken three hundred times a day; whether you're going down to the Underground or walking down the street, there are cameras everywhere. New York does not have so many yet, but they're coming as fast as we can put them in. I think the general rule is that crime has been falling here more substantially than anywhere else in the United States for the last fifteen years. And miraculously it has continued to fall and I think will continue ever on, partly because it's a great city and because of the density, but in part it's because of the cameras.

There have been two high profile murders of young women in this city in the last six months. One, a woman emerging from a bar down in the Bowery at four o'clock in the morning, whose body was later found out in the outer edges of Brooklyn, and another who was at a bar in Chelsea. She was found the next day in a cheap hotel in New Jersey. In both cases, the murders were apprehended within a matter of a few days, because of this new technology. They were caught on video cameras that were outside just picking up things on the streets. Anyway, I'm predicting the crime rate's going to continue to fall.

Fourth, I think there's a new respect for government employees. It's often said that New York respects only money and only the Masters of the Universe. But just think of the feeling about firefighters after the World Trade Center. Women talk about how their jaws practically hang open when they see fire trucks going down the street. There is intense respect for those people who ran into the buildings when everybody else was running out. And it's not just firefighters.

I think what's also remarkable is, as far as I know, not one person did what I would've done, which is to say, "You know, I'm not making enough money to go into that building, I quit right now." Indeed there were almost sixty firefighters who were not even on duty who died at the World Trade Center. Sal Casano, who is the Chief of Operations for the Fire Department, told me that the time of the attack was bad for the fire department, but good for the city. Good for the city in the sense that it happened at 8:46, 8:47 in the morning. New York is a late go to work town. Most of the people who worked in the World Trade Center in fact weren't there. Maybe because the Jets won Monday Night Football, maybe because of voting in the primary, mostly people just don't go to work at that time.

But for the fire department, there was a shift change at nine that morning. So as those alarms were ringing all over the city, those firemen who were going off duty at nine o'clock could have just said, "I'm outta here. I'm headed back to Rockland County or wherever. I'm going." But instead they stayed, and as I said, sixty of them died.

Another editorial is that Giuliani, who was mentioned earlier today, has generally received high marks for his performance in September 11 and immediately afterwards. We forget that his reputation was sinking in New York City before then. To me it is just the reverse. I think he was a very important and effective mayor before September 11, maybe well before. The turn around of the city can't be disassociated from Giuliani, whatever you think of him. But I think his performance on 9/11, or with things having to do with 9/11, was not good. The fact that the fire department could not communicate with itself ten stories up, has got to be his fault; he was Mayor for seven years before that. The fact that the police department and the fire department couldn't and didn't communicate because their radios were on different frequencies. . . you know, somebody has to take the blame for that.

I have been saying since 9/11 that there will be no more terrorist attacks in New York or in the United States. I realize I'm almost alone in believing that, but I think this country is remarkably effective once it gets itself together in organizing against internal dissent.

I think New York's place in the world is good. It's still glamorous, it's still sophisticated, it's still exciting, it can still claim, as no other city anywhere can, to be the greatest city in the world.

But the United States, especially its political leadership, has not risen in the estimation of the world; quite the reverse. Once we were respected and admired. I'm not saying around the globe, but in many parts of it. But now if you travel internationally, you see that we're so often regarded as bullies, as hypocrites, as fools, as embarrassments. Some of you may have seen that on the polls that list that the greatest threats to world peace—a list that includes Al Qaeda and Osama Bin Laden—the United States comes out ahead.

This thinking is true not just in Muslim countries, but also in our major western European allies. Even in those countries that cheered previous administrations or cheered Americans or loved us for saving them no longer feel respect and admiration for us.

Obviously, September 11 is still a horrible day for this city. From this campus, to look out and not see those buildings . . . I was one who didn't like the buildings but it was only after that they were gone that I realized we needed them. If you're coming back on the New Jersey turnpike, they were the first things you would see and you knew you were almost home. If you were lost somewhere in Brooklyn, you could look around and see the World Trade Center and get your bearings.

I think the city generally has responded with distinction. In comparison with New Orleans, we can see there are many differences. New Orleans was a bigger tragedy, at least in terms of the physical city—it was a whole city. But I don't think the uniformed workers or the administration on any level in New Orleans distinguished themselves. New York comes off well in comparison.

Like the rest of you, I hope that our fall from grace in the world will be temporary, that the United States will once again be a beacon of freedom and liberty for all the peoples of the earth.

Thank you very much.

BEVERLY KAHN: This panel is entitled, "America's Place in the World." As such, there are a number of things we need to define. What is America? How do we define and characterize the world today? We've had some people, like Mr. Friedman of course, saying the world is flat and so forth, but how do you characterize the world today? I'll ask Steve first. If we're trying to define what America's place is or ought to be, what, to your mind, is the world today?

STEVE CLEMONS: That's one of those big cosmic questions on which one could write a major thesis, and I won't. I think that the bottom line on the question of, "who are we and who are they," is one that, before 9/11, was becoming blurrier and blurrier. The borders were in decline, the world was becoming more integrated, we increasingly had people, institutions, money, and ideas moving across borders. We had a kind of globalization that was characterized as a high trust globalization.

Today the definition between who they are and who we are is much starker. We are now back to states. Borders are becoming thicker. People and ideas and money and institutions are moving less freely across borders. I think there are costs to that. That kind of globalization is a messier globalization.

It also challenges us to think twenty, thirty, forty years from now about what the world is and what our place is in it. And we must do so in a way that we haven't had to since the end of World War II—and I'm not satisfied that the United States is having that discussion. I think the other remarkable thing is that it's harder to imagine a stronger position than the United States had vis-à-vis the rest of the world than we had then on September 11 and immediately thereafter. The mystique about America and America's place in the world was very, very sound.

Today I would say, for a variety of reasons, that the mystique about America's place in the world has been shattered and punctured. Our allies won't count on us as much as they once did and our foes are moving their agendas. America's position overall is weaker. And it's not weaker because of 9/11. September 11 actually strengthened our hand. I think our current situation forces us to come back and reconsider many of these questions about that.

I don't know if that definition fits, but I do think that those are the contours that matter.

BEVERLY KAHN: Would you suggest that it was 9/11 that instigated the stark identities and the enhanced conflict or tension?

STEVE CLEMONS: No, I think that 9/11 instigated an incredible amount of empathy from the rest of the world, and also hope. Whenever you go around the world and talk to people in India or Kuala Lumpur or Pakistan or Africa, as I have, one of the things that's always characteristic is how shocked they are at how boldly America walks through the world doing whatever it would like to do without a sense of constraint. These people are all too aware of the constraints on themselves and their nation. And they are very envious, and they aspire to that kind of thing. So there was a great deal of empathy and good will.

What I would argue, and I don't want to make this political, is that we made decisions down the road that were perceived to have less legitimacy in the eyes of many parts of the world. After the Afghanistan invasion, we began to push the world over the line in terms of what it could stomach, of what it felt was legitimate global action. That collapse of legitimacy has also meant a collapse of American leverage in the world towards its great purposes.

BEVERLY KAHN: America, Ken says, is perceived as a hypocrite and bully, no longer a legitimate leader. Alice?

ALICE GREENWALD: I come at this a little bit differently, but I want to say I think 9/11 happened to people as much as it happened to a city and a nation. It happened to people, like you and me, who got up in the morning and got their cup of coffee and

192

went to work doing what they did everyday, and they got caught in the vortex of global terrorism.

Most of us live our lives, not just those in the United States but around the world, doing what we do every day and not thinking about how the work we do or the activities we engage in or the products that we help create actually function within a global network. What 9/11 did to all of us was communicate a kind of democratization of vulnerability that America did not have a sense of before. September 11 changed a sense of who we are as an invulnerable place. That terrorism could happen else-where, we understood, but we never believed it would happen here. That was a fundamental shift of consciousness. And not only for Americans, but for others outside of America. We became like everyone else. I will share a story.

I was at the Holocaust Museum on 9/11. A very dear gentle-man who was a holocaust survivor came in and kept saying, "the impossible has become possible." That's all . . . that's the only way he could phrase it. For him it was impossible to think that the heaven that he had found in America after World War II was now under attack and that people were being murdered in this indiscriminate way. That was not warfare as he knew it; it was individuals going about their business being slaughtered.

The second thing I want to say is that being a student of the twentieth century, because of the work that I did at the Holocaust Museum, I have come to the realization that much of the twentieth century was about the perpetual creation of refugees. It happened after World War I, it happened during World War II and after World War II. At the end of the twenti-eth century, we saw the breakdown of national borders with the disintegration of Yugoslavia and the resulting genocide that occurred there. If the twentieth century was a century of refugees and the breaking down of national borders, the chal-lenge that we have in the twenty-first century is that we don't have new borders. We have pan-nationalism at war with differ-ent groups of pan-nationalisms. That's a new kind of structure that we have to confront and begin to understand.

BEVERLY KAHN: You've focused on what happened to people, and the sense of psyche and the sense of vulnerability. But you've also given us some different definitions of the world. Experiencing global terrorism, vulnerability, increased interdependence that leads to vulnerability.

Let us move on. Nikki, how would you define the world today, how do you perceive it? You certainly have a personal perspective on this.

NIKKI STERN: I want to get back to something Steve said, which is that many people in the countries he visited were almost envious of the way we go boldly into the world. I've been looking at the question of moral authority. I started from a very personal standpoint about the way we assign moral authority in this country, from people who are victims to celebrities, and I logically moved up to moral authority in the government.

This country has always seen itself as uniquely blessed. That's been true for our entire history. And indeed it is. Yet, this has led to this assumption that because we lead good and just lives, that whatever we do in the world is good, that our actions are justified. But the problem is there are times when they are not. Everything is backed to high and implied moral authority.

I finished reading Madeline Albright's book and I was struck by a comment that she made in her book *The Mighty and the All Mighty*. She said history would be far different if we did not hear God most clearly when we think he is telling us exactly what it is we want to hear. I thought that was amazing. My father used to tell a story about a sign he saw in a window, that said, "God we trust, all others pay cash."

We define the world in terms of good and evil. We have leaders who say, "I hope people approve of what I'm doing, but if they don't I'm going to do it anyway, because I answer to a higher authority."

It occurs to me that maybe a better description would be, "trusts God . . . does not play well with others." That concerns me, that really concerns me. The other thing about the world that I wanted to address is that in the countries that I've visited—and I've dealt a lot with young people from Muslim countries—I've found that a lot of people actually still like Americans. They really like us as individuals. They are impatient, at times, with our ignorance or prejudices against other cultures. But they like our outgoing, supposed optimism, our candid attitude. But they are furious at these elements of our foreign policy, or what they perceive as our foreign policy, that have overwhelmed how they used to see us, when our "can do" attitude was in service of an agreed upon greater good, such as working against genocide or starvation.

By not looking at how we apply our moral authority, by just saying, "Hey, we're good, so anything we do is good . . . it's gotta be good, because we're good and we're democratic," we've lost our way. And in a way, we've actually lost our moral compass.

BEVERLY KAHN: Nikki, thank you for making the distinction between Americans as individuals and America in terms of what its leaders, what its government, and what its policy represent. One might also add what our businesses and our corporations represent, which may yet have a different definition. Ken, you began by saying that America has not risen in respect while New York has. And Steve said that America has less legitimacy and Alice raised questions of moral leadership, and certainly Nikki did too. You all have raised questions of value, of what is good and evil, and what America does stand for.

America once stood for liberty and opportunity and helping others and caring. I remember my civics classes. That's what America stood for. That's what the Statue of Liberty stood for. Yet, you all seem to be echoing the notion that America's leadership, its moral authority, and the values it represents are now unclear or maybe disdainful to people in the world.

KEN JACKSON: I agree with Nikki in that I wonder if we have lost our moral authority, but—and I'm just going to be outrageous here—sometimes I think President Clinton would have been more forgiven if he had just asked forgiveness, if he had said, "You know I was stupid. I did the wrong thing." And I wonder if the world wouldn't respect us more if we said, "Yes we believe in liberty and justice. And we also believe in cheap oil.

And we're just going to have to have cheap oil and a dependable supply and therefore, get out of our way."

I've found in talking to international people that somehow they almost feel like we're hypocrites, partly because we won't at least say why we're doing what we're doing. Which is that it's about oil, as much as it is about anything else. You know Kuwait had nothing to do with freedom and democracy and neither does Saudi Arabia. It's about oil. And maybe if we just come clean about it, and say, "In order to have freedom and democracy we feel like we need oil, and therefore we're willing to kill anybody in order to get it."

Then maybe we would enjoy a little more respect. Maybe we need to say, "Hey, this is our weakness. We built a culture around inexpensive transport and shopping malls and big yards and easy travel and we can't go back on that." I would say that's a historical development. History's important here. We can't live without oil in this country, at this moment. Therefore, anybody who challenges that is going to run up against our entire era. Maybe we should just be more honest about it.

BEVERLY KAHN: Some members of the audience agree that we are not honest and that we're hypocrites and we need to be candid. Do any of my panelists have reactions to those notions? Is it a reaction to that question?

NIKKI STERN: It's a little outrageous, Ken.

KEN JACKSON: It's towards the end of the conference, you know.

NIKKI STERN: I think there's a mix. I honestly believe, and I always feel that I need to say this, that this is a great country. But at the same time, because it is a great country, I hold it to a higher standard. I think that everything that people do as individuals, that countries do, that governments do is and should be a confluence of what is in the best interest of the country they serve and also, hopefully, works is in the best interest of the world. I might be an optimist but I believe there is a way to combine both things. Obviously, it would be in our best interest to be less dependent on oil, but if we are dependent on oil there could be a way that we could use oil and produce jobs for people and it might require a little more honesty or directness. The other thing I want to point out, is that hypocrisy is not unique to the United States.

I am always fascinated by talking to young people from incredibly privileged backgrounds in countries where the wealth is disproportionately controlled by a small group of people. When I would mention the concept of cleaning up their own backyard, they would blame the United States, saying it was our fault for supporting the regime that kept them comfortable.

There is an Arab-American comic who has a joke about this: You know, if you don't have any food, you blame Israel and the United States; you don't have any grain you blame Israel and the United States; you don't have sex with your wife you blame Israel and the United States. I mean there is a point in which the hypocrisy works both ways. But that has always worked in service of the way governments advance their agendas. I'm not certain how you get around that.

KEN JACKSON: If I could quickly respond?

BEVERLY KAHN: Yes, of course.

KEN JACKSON: The oil issue is complicated. You think about who's providing this oil internationally, you tick through the countries: Venezuela, what a mess; Saudi Arabia, a government that has been sticking its neck out to try to solve the Israeli-Palestinian issue. I know it's hard to consider the Saudis moderates, but nonetheless, we've just had an example where a Mr. Nasrala has just secured three hundred prisoner releases. So, you have moderate Arab governments that are oil suppliers that are now not looking so good to some of their own constituents. You have Iran, obviously brewing problems. And Iran is neck-and-neck with Saudi Arabia in terms of the oil and natural gas supplies that it might bring online. And yet, we are tripping towards a hot, invasive action towards Iran. You can almost assume that down the road there will be a nuclear program in

Iran, an Iran that tilts towards the other biggest natural gas and oil supplier in the world, Russia.

So, it's one thing to basically admit that you've got dependencies, it's another thing to construct U.S. foreign policy in a strategy that doesn't just acquiesce to all these realities, that does talk about America's great purposes and does talk about hope and opportunity, but also is realistic and doesn't undermine the interest of the nation or our ability to achieve these ends. I think that beyond the question of hypocrisy there is a legitimate need for debate and discussion. You have a luncheon speaker, Bill Kristol, who I respect and admire. Though we are very different, I have learned a lot from him. I think my team is beginning to score some points from learning some lessons from Mr. Kristol.

I would point out to you the President's speech on Wednesday. Embedded in the President's speech was a very, very important revelation. That ongoing in the Administration—in the bowels of the White House and government—has been a very divisive debate among those who felt that after 9/11, the White House needed a kind of executive authority that was exceptional. That we needed to suspend our typical system of checks and balances in this country, that we needed to curtail our protection of civil liberties that were provided to all, even our enemies, because of the unique nature of these times.

Yet, the President gave a speech in which he basically indicated that that war paradigm, which had been created in the early days after September 11, is over. He said very clearly, "This is a nation of law. You see these black sites? These CIA detention centers? They are going to be emptied." This is a major step towards returning the United States back to the system of legal norms and transparency that has been missing.

Why would the President do that five years after 9/11 occurred? It's an indication, in fact what most of the world thought, that we had suspended some of the behaviors that made us most admired. I think that is not necessarily hypocrisy, it's a sign of a genuine debate among good people who see these things differently.

BEVERLY KAHN: Let me follow up on that. Do you think that the U.S. in recent years has been so focused on, and some could say distracted by or obsessed with, terrorism in Iraq, that we have not been playing a leadership role in the world and solving real problems beyond terrorism, including dependence on oil, environment challenges, the problem of AIDS around the world, and poverty. Have we been so focused, so obsessed with terrorism that we have lost some moral leadership because we're not addressing real world problems?

STEVE CLEMONS: I don't think it's an easy either/or. I opposed the war from the start, but I wrote a *New York Times* article where I said, "Okay, I'll stop bitching about Iraq. How do you get an occupation right? How do you basically get things in gear?" And I suggested something like the Alaska Permanent Fund for Iraq where there would be a class of economic winners created in Iraq who would be less dependent on their local mosques and see us as helping to deliver resources.

What if we had gone with some notion of Condi Rice's transformation diplomacy rhetoric, and had gone in and said the Middle East needs to be brought into the modern world and Bin Laden is not just a terrorist. Bin Laden is a performer trying to reach an audience and trying to perform to that audience and look legitimate in their eyes by exploiting grievances, in a very cynical way. What are those grievances?

America should be stealing that audience from Osama Bin Laden. That is the way to beat terrorism. You do that with economic programs, you do it with exchanges, schools, lots of kinds of things. But we went in on the cheap. We thought Iraq would be Romania. That we would roll through the streets and be cheered and that everything would be easy and we would trip through there to Iran and we would establish a domino theory of removing tyrannical governments around the world and that it would be a low-cost venture.

The fact is, it's not. We need a new Marshall Plan, with Marshall Plan strategies to help provide resources and opportunities and networks. I think we, as a nation, weren't prepared to do this. We went to war on the cheap, and we've basically been doing the recovery on the cheap. There are other choices.

I don't believe it's Iraq versus global warming or Iraq versus AIDS in Africa. I think that we need to develop a new set of strategies that takes the best parts of this country and brings them to the world in a very consistent coherent way. But tit-for-tat or this-versus-that is not enough. We need a new paradigm for thinking about these things.

BEVERLY KAHN: And you suggested that, in your mind, the President's speech suggested that he's had an epiphany?

STEVE CLEMONS: I think the President's speech was an important first step saying that off-the-books black sites, secret detention centers, kidnappings on the streets of other countries are not consistent with American democracy. I mean, I remember very well during the Cold War, that that's what the Soviets did. Nikki talked about moral credibility, the savaged moral credibility of the United States. I think the President realized that there had to be an exit out of that game. In fact, if you read the *Washington Post* today, even the CIA allied with the State Department and said it wanted out of this black site business and wanted to come back into a system of predictable law and processes.

Many people won't like what the President suggested, but in my view it's nonetheless a very important step in the right direction.

BEVERLY KAHN: Regaining moral credibility, legitimacy. Nikki?

NIKKI STERN: I do agree with you that, and I think that I was trying to say this before, that there is a balance. And I will reiterate this: There is a confluence, a point at which what you need to accomplish in the best interest of your citizens balances against whatever somebody's definition is of what is morally right. I also never had any doubt that the people in the Executive Branch believed, they believed sincerely, at least some of them did, that their approach, "knock a domino, boom,

boom, boom, boom, boom, what a piece of cake and we'll be home tomorrow, mission accomplished" would work. I never doubted that they believed in their heart of hearts that this was the right thing to do and the right way to do it.

My point is this: When you open up the possibility that more people can get involved, hopefully you also open up the possibility that more experts can get involved. And while experts don't know everything, their experience is an important contribution. What ticks me off to no end is that there came a point in time when the Executive Branch absolutely needed as wide a variety of expertise as possible, and instead they put cotton in their ears. My concern now is whether we have now trained ourselves, at least from a foreign policy and government standpoint, not to listen. My first reaction, even though part of me agrees with what you're saying, Steve, about the appropriate steps, is too little, too late. Because, we have to go forward now and we have to do something now.

STEVE CLEMONS: Part of me thinks that too.

NIKKI STERN: We have a serious PR problem and we have a serious listening deficiency and it's not good enough to go to people that might be able to provide some perspective and some help with your hat in your hand after the fact. You've got to build a just, humane, workable role for America in world affairs that relies on a variety of opinions and experiences.

BEVERLY KAHN: Ken, do you want to follow up on that?

KEN JACKSON: I want to ask a question. I want to ask whether we're at war. I mean, you couldn't tell it from my life and I suspect from most people's lives. In other words, though I think the war analogy is constantly thrown out there, I wonder if we can really say that we're confronting a unique enemy. I wonder if it's a fair analogy to compare this to Hitler. After all, wouldn't you say that Hitler and Stalin could be terrorists with the best of them. And yet they also had, especially Hitler, a war machine, the likes of which the world had never seen. I think that saying that Osama rises as a threat to the United States in the same way that Hitler and Imperial Japan did, strikes me as twisting history just a shade, to exaggerate and in a sense to make our own responses more eloquent. Sure, Abraham Lincoln suspended habeus corpus during the Civil War. This country's survival was at risk, really at risk. I question whether we really are in a global war. The terrorists want to kill us, but I wonder whether this giant nation is giving some legitimacy to people who otherwise don't cut much of a figure.

We compare 9/11 to Pearl Harbor, but nothing's happened since 9/11. Look what happened after Pearl Harbor: Wake Island and Guadalcanal and Tarawa and Iwo Jima and Okinawa and North Africa and Sicily and D-Day and the Battle of the Bulge. There was a lot more. I wonder if we are really worthy of our grandparents, to shiver in our boots because some terrorists have done a little bit. They aren't terrorists on the level of people earlier in the century. At least they haven't proved it yet. Maybe they will. But I don't think the evidence suggests that the power behind those regimes, and let's call them all illegitimate, Hitler as well, is the same thing as that of Nazi Germany.

BEVERLY KAHN: And indeed, on our opening session two days ago, David Gergen suggested that we should stop using the word war, and "war on terrorism." So he agrees with you.

What do you think Alice? Is this a war? Are we fighting a war? Who is the enemy? Can you define the enemy for me? And what's the goal?

ALICE GREENWALD: Again, I don't feel like I am, at all, qualified to do that.

BEVERLY KAHN: But democracy depends on citizen involvement and action. So what's a citizen?

ALICE GREENWALD: Actually, what I would like to do is go back to the moral compass question. I think that is really key here. You posed a statement earlier, that what America stands for is liberty and freedom. You know, I work at One Liberty Plaza, on Liberty Street. I go to Liberty Dry Cleaners. I eat at Liberty View Chinese Restaurant. I don't know what liberty means anymore. It has become absented of meaning. And I worry that it is used too loosely. Freedom is thrown around like we all know what we're talking about. I don't think most Americans know what freedom is. I don't think we value it; I don't think we display advocacy of it in the world. And to be a moral compass, you can't just use the words, you have to follow through with the actions.

After World War II, in 1947, the United States signed on to a genocide convention. We signed onto a convention that we today don't honor. There is an example of a moral compass way off kilter.

I had the privilege this spring of being at the commencement exercise at NYU, where Justice Anthony Kennedy was the speaker. Normally you go to these things and you expect to yawn your way through very packaged comments. To my utter amazement, I was riveted. This man stood up in front of 20,000 graduates and was angry. He was visibly angry. This great Supreme Court Justice was seething as he spoke to these young people. He said, "We have let you down. My generation has let you down. We are leaving you with responsibility for a world that is a mess. And I don't have words of wisdom for you. The only thing I can tell you is that we have got to stop looking at the other people around the world as 'other' and start thinking about them as husbands and wives and sons and daughters and cousins and colleagues, we have to start thinking that these are people you will have to deal with as you go out into this world. They don't look at us this way, but they're going to have to. We have to do the same in reverse."

We're equally responsible for the world we live in. I took that to heart. I think it is really a key message for this century, going forward. So when we ask, what does America stand for? I'm not sure any more. I'm the daughter of a World War II veteran who marched in every Memorial Day parade. He was President of the American Legion, President of the Jewish War Veterans in our town, and still I don't know what it means, anymore, to honor American values, because I don't see them demonstrated. If we as a generation have an obligation now to give to our children a world where America can be responsible, we've got to understand the world we live in and that it's populated by human beings, not just others.

BEVERLY KAHN: And you said you didn't have anything to say.

ALICE GREENWALD: I'm sorry.

BEVERLY KAHN: What democracy requires is for people to get angry, for citizens to be public intellectuals, to speak out and help right our moral compass which is off kilter these days. Charles Dickens begins *The Tale of Two Cities* with the words, "These are the best of times, and the worst of times." My pen would write, "These are the worst of times." When I look around, I see that we are hopelessly dependent on foreign sources of energy. I see the proliferation of nuclear weapons to increasingly irresponsible regimes. I see that we are failing to address serious problems of global environmental challenges. That individual freedom and privacy are threatened. That we are seemingly unable to care for our own, especially as witnessed in the aftermath of Katrina. I see the extent that race and poverty still determine one's fate in this country. We seem to accept layoffs and cutbacks in health insurance and growing economic inequality as part of the fabric of our times, believing there's nothing we can do about it. Instead of welcoming immigrants, we're now talking about building a fence to keep them out, one that would rival the Great Wall of China. We've even begun to doubt whether our own elections are run fairly. These *are* the worst of times. We are in crisis, we need public intellectuals, who are angry, to speak out and speak what is democracy. Is it toppling regimes? What was democracy, Ken, in New York City when the immigrants came? What is democracy? What do *we* represent? Can you help right our compass or build a new compass in this new twenty-first century?

KEN JACKSON: I don't know if I can answer that, but I will say that there was a famous ward boss in New York, Big Tim Sullivan of the Lower East Side. He and George Washington Plunkett would talk about how the new immigrants from Ireland or Italy, who had never voted in their life, would get off the boat and three or four days later they'd be voting.

By the way, technically the law said they couldn't vote, but this was Tammany Hall in New York. I do think that in that sense a democracy was introduced to a lot of people very early on in New York who had never really seen it at all working. Was it perfect? Of course not. But nevertheless, their opinions were solicited and sought and they were able to make a hugely important impact.

The Ward bosses offered them help with a job and help with the law and help with a lot of other things. But they gave them the dignity of respect. They said, "We'll do this for you, but we need something from you, too. We need you to vote (early and often as it were). We need you to get out there." That allowed a person to keep his dignity. I think this city did have an important role in democratizing and Americanizing young people from other countries. I think it's still the great glory of New York that it still welcomes people. The city has done a better job of welcoming immigrants than any city in the history of the earth. I think if we were really to close our borders in a serious way, this city would suffer more than any others. Because we really depend on their energy and aspiration.

BEVERLY KAHN: Thank you. Perhaps it was bringing them into the voting process, to the community process, that was democracy's role; perhaps it's more than procedures, more than free elections.

KEN JACKSON: By the way, I think the casualties in the World Trade Center, in many ways, reflect the diversity of New York City. I mean could you have gone anywhere else and hit a single building and killed people who represent so many different religions and groups and races and ethnicities?

BEVERLY KAHN: Ninety-two countries. Doris Kearns Goodwin talked about democracy as shared sacrifice, as working together for the common good, for the good of the community, civic engagement and higher goals and values, equality, individual rights and privacies, and so forth, liberty.

Steve, would you share your reflection on helping us set our moral compass?

STEVE CLEMONS: I'm a bit worried about this country, because I think people need to wake up and I don't know how to engineer that. I think things happen organically. You know, during the Cold War when I was a young guy at UCLA working at the Rand Corporation, I used to hang out with the Soviet Watchers, the high-priests of Soviet Affairs, who were the biggest "strutters" in the foreign policy business. This was a small group of people who were controlling and directing America's engagement with strategic weapons, through the U.S.-Soviet relationship and the American public indirectly, through leaders that had delegated to this small group the responsibility to do this. Because they thought there existed in the United States a general consensus about how the world was shaped. I think that's over and I don't think the taxicab driver anymore can give you the quick picture about Mutual Assured Destruction or the Cold War or the bi-polar world. We're back to a point where we need to get the engagement back so that we can come back to a consensus of what we are doing.

I don't think we're anywhere near that process. I, too, hate the metaphor of the "global war on terror." Because I think it is a false metaphor that has created an imprecise way to respond to these things. Like Ken said, during World War II this great nation fought two wars on different sides of the earth. And here, five years after 9/11, we still haven't solved the problem of stealing Bin Laden's audience away. And frankly, it's not a war against Bin Laden. That morphed the day we began sending those forward troops to Afghanistan and Iraq.

It morphed into something much bigger. The question then is, if this isn't a war—and I'd hate to see the state of the U.S. military when we were at war, because it's hard to imagine the

U.S. military more broken or more on the edge of complete fatigue and breakdown than it is today—then what is it we're involved in? I also think that the question about shared sacrifice that Doris Kearns Goodwin mentioned this morning is a good one. I don't think most Americans, unless they know someone in the military or National Guard, someone whose financial situation has been undermined and wrecked, a family that has lost its healthcare because their National Guard husband or wife has been deployed, share the sacrifice. There's a sort of subterranean shock going on that most members of Congress don't feel.

Max Baucus does, from his nephew passing away. Duncan Hunter has a son that's deployed. But there are very few members of Congress who have family members in the military. This is also a reflection of the segment of our society that has direct contact with the sacrifices going on. It's not being broadly shared.

BEVERLY KAHN: I think that some of you in the audience have some questions. We're going to turn to you and try to respond.

AUDIENCE: Thank you for the comments from the panelists. My name is Tim Davis. I'm CIO for the King Capital Institute here in New York. I was raised in Florida by a loving mother and was taught to remain diligent in all my efforts. Doing so brought me to the place where I was actually able to graduate from college. Going into the Air Force, I became a navigator pilot. After I got a medical separation, I went to Harvard.

I appreciate the comments from the audience and I really appreciate the sentiments expressed in regards to the American culture and how we have progressed. I believe that yes, we've had some bad days and also some good days. Some successes and some failures. However, I believe we've had more successes than failures and also I believe we have come far and yet have far to go.

Calvin Coolidge indicated that prosperity cannot be separated from humanity. And of course, Albert Einstein was correct in the assessment that great spirits have always encountered violent opposition from mediocre minds. I believe that as a culture, we should focus more on blending the public and the private sectors. I would like to ask the panelists how do we encourage the public and private sectors—the federal government, and the farmlands and tribal lands—to harness their energies, to harness all available resources to encourage the American people to influence the rest of the world.

The reason I say that is because they look at us from a mirror image. They want to see our leadership by example. If we can remain steadfast and provide leadership by example, I believe we'd be able to see the results that we hope for from the rest of the world. That's my question.

BEVERLY KAHN: The question is, how can we provide that leadership by example. How can we as individual citizens or as members of our government or a public/private partnership regain that moral authority and respect? I think there has been a consensus among the panelists that we've lost some of that legitimacy and respect in the world. Our role in the world is therefore diminished. How do we reassume that?

STEVE CLEMONS: Let me say just one quick thing. When I was in Political Science, and I don't know how many of you are Political Science students here, if you wanted to study a social system, whether it be another country or nation, or your local city council, you don't pay attention to it on a day when everything is going swimmingly. You pay attention to how it operates under stress. It's only when you see a system under stress that you really know what the norms are, what the real operating procedures are.

I think if we really want to restore American moral credibility in the world, we must show the world that we are comfortable with our system of law, our system of justice, our system of habeas corpus, our system of appeals, our system of checks and balances, and our rough and tough form of democracy. And that we don't suspend those systems the moment that we're shocked or surprised or caught off guard. Our strength as a democracy is to show that we have confidence in what we're all about on the worst day. That's what the world wants to see. If we get back to that, which I'm hopeful we may be tilting towards, that's the best thing we can do.

BEVERLY KAHN: Thank you. Any other comment or follow up on that? If not, I'll take another question. Back there.

AUDIENCE: I'm Michael Kramer; I'm a downtown resident. In the past five years there has been a spread of mass communication devices, televisions, cell phones, computers. What do you think of the role that the media, both inside and outside of our country, has played?

BEVERLY KAHN: Because Nikki in her opening comments talked about marketing democracy, I'm going to ask her to take it, Nikki?

NIKKI STERN: Oh my God, I'm so infuriated at media manipulation. I think we live in a 24/7 world. And what I think is singularly appalling is that while the number of outlets that are available for people to express themselves has increased, all of the information that we're receiving is broad, but not deep. And there's huge repetition across the spectrum. I think it was Alice who was pointing out this morning that the most important news item was about a suicide bombing in Kabul and then . . .

ALICE GREENWALD: And it got equal time with Lindsay Lohan getting her pocketbook stolen in Heathrow.

AUDIENCE: It infuriates me. It's just another blip on the screen; it just goes in your head, you know you don't pay attention. We're not asked to pay attention. As Doris Kearns Goodwin said this morning, we live in a celebrity culture. Right now I think it's a global phenomenon. It's made us, I think, stupid.

NIKKI STERN: I absolutely agree, it really fries me. I'm working on a treatise on moral authority and why we in this country seem to give it away so easily.

But I was talking to someone the other day and I said, "Why is it that we look to others for moral authority? What is wrong with people in this country? We're educated; we're not that stupid. I mean people can talk to you about stock markets, they can talk to you about box office receipts, they can talk to you about baseball stats, they know the things they know. So why do we give this away?"

He said, "You know, people don't have time." My response to that is, "make the time." We're talking democracy, we're talking liberty, we're talking freedom, we're talking the United States. Make the time. In conjunction to that, we need the media to help, not hurt, we need the media not to treat us like we're stupid, and we need not to buy into the fact that we're stupid. I don't know how to bring another JFK or RFK on board to make voting and public service sexy, like it was for me when I was in junior high school. But we need to make the time to look at the decisions that our leaders are making and analyze them. And whatever you make and however we decide to vote, that's fine. We do have the time and the media's colluding in not bringing those issues to the floor.

BEVERLY KAHN: I'm sorry for interrupting. But Steve has to run off to be interviewed on CNN.

STEVE CLEMONS: On 9/11.

BEVERLY KAHN: So we have to say goodbye to Steve, but he has a blog.

STEVE CLEMONS: I'll blog about all of you.

BEVERLY KAHN: Thank you very much. We'll look for you on the mass media, the informed mass media. I want to thank all of our panelists and audience members for a superb discussion. Thank you again for participating.

A CONVERSATION WITH MARK SCHAMING, CURATOR, NEW YORK STATE MUSEUM

Where were you on 9/11/2001?

I was in Los Angeles on September 11, about to fly to New York out of LAX. I remember seeing the second plane hit the South Tower on live television. I flew back to New York when they reopened LAX later that week. It was a memorable flight, one I'd never want to repeat. It's a good question—everyone knows where they were that day.

In your panel discussion at the conference, you mentioned the recovery effort at Fresh Kills was like covering "history as it was happening." Did you come to this realization immediately after the attacks, or was it not until you were actually on site? How would you describe the importance of preserving these artifacts?

For me, it all seemed surreal then historic. I was at the WTC site twice in the fall of 2001, and it was overwhelming. A week later, I saw the Fresh Kills recovery operation for the first time. It was clear that the police and FBI knew they were in the middle of a historic event. There was this intensity about the days and a sense of disbelief. The anticipation of something worse to come hung in the air. Fresh Kills was unlike anything we had seen, and we felt it was like a hidden history. Every day spent there was the same in some ways, but we saw more with every visit. From the first day at Fresh Kills, when we saw very few recognizable objects in the sorting fields, we knew it was of utmost importance to save something tangible. Every object we saw seemed historic and the stories were memorable.

Your permanent exhibit, titled The World Trade Center: Rescue Recovery Response, *is located at the New York State Museum in Albany. Is there a particular artifact or portion of the exhibit that resonates most deeply with you? Why?*

There are many pieces in the museum that are especially important to me, but three things in particular come to mind. We collected an airline seat belt fragment recovered at Fresh Kills.
You can't tell from which plane, which passenger's seat, or
whether it was the seat of a flight attendant, traveler, or terrorist. The object resonates with the day. The Engine 6 (pumper)
truck is a powerful and dramatic artifact that holds tremendous
stories. For me, it's the voices of the FDNY firefighters who died
and of the people, especially FDNY firefighter Billy Green, who
lived to tell us what happened that day. This will be even more
important as time goes on. The small objects of everyday life are
also particularly compelling. We saved one broken telephone
handset. They found only two in ten months at Fresh Kills in 1.9
million tons of debris. You can't look at it without thinking
about who might have spoken into the phone, whether they
lived or not. The piece hints at the extent of the destruction. The
Recovery section of the exhibition resonates most deeply with
me, as I spent so many days at Fresh Kills.

*You have close relationships with many of the emergency service
providers in Lower Manhattan, especially Engine Company No.
6 on Beekman Street. How has this contributed to the work that
you do on this very sensitive subject matter? Has it made the
work more difficult for you personally?*

Without the relationships we forged with the FDNY, Engine 6
in particular, there would be a much more distant interpreta-
tion of the collections the museum saved. In the end, the human
voice is what makes an object important. Well before we collected anything, we developed relationships (which mostly became
friendships) with the NYPD, FBI, Port Authority, FDNY, and
relief agency personnel. I think they saw us as people whom
they could entrust with their stories, and then the artifacts.
This becomes historical record. We worked alongside them, ate
together, talked a lot, endured their unrelenting humor, attended funerals, went to parties, and shared everyday life. Through
this, we came to know the things that will never go into history

books, things they don't talk about at home. You don't just arrive, take pictures, and collect things. We developed a clear sense of the texture of their lives, jobs, September 11, and the aftermath. For me, it's made working on this project that much more rewarding, and given me a real perspective about the difficulty one feels about the tragedy. I still think about it every day and feel humbled by what these men and women are living through.

Many of the Aftershock participants felt that the exhibit was the most powerful and compelling portion of the conference. Do you think the exhibit's impact was enhanced in the larger context of the panel discussions and programming, or are these the kind of objects that speak for themselves? Was there anything particularly new or moving that you took away from the proceedings?

The sessions were truly memorable. Overall, I was reminded about how close we still are to the events of 2001, and in many ways how the issues are still so critical. History is still being written. The Pace exhibit installation was unique for a few important reasons. The proximity of Pace to the WTC site and the Engine 6 station was important, and the presence of the speakers and the attendees created an incredible synergy with the exhibit. It was moving to see, as a session ended, people in deep discussion leave the theater and move into the exhibition. The artifacts on exhibit were touched by history. It's impossible to see these artifacts and not connect in a very visceral way, but they were far more meaningful in the presence of the people who spoke and of those who came to hear them.

A CONVERSATION WITH TOM FARKAS, EXECUTIVE PRODUCER, NY1

Where were you on 9/11/2001?

I started out the day in Katonah, New York, in a dentist's chair, listening to one of those easy listening stations. When they announced in their very low key way a report of a plane flying into the World Trade Center, I knew it was time to go. By the time I got to my car and switched on WCBS-AM, they were talking about a second plane. I made my way towards Manhattan, and as I got to the Major Deegan expressway in Yonkers, I was able to see the smoke rising from the buildings

What was the immediate reaction among you and your colleagues to the 9/11 attacks? Was there time to formulate your own ideas for perspective and content, or was there a more pressing need to simply document what was going on in the city?

When I arrived at the station, we were already deployed in most every direction. I grabbed a camera and drove a reporter, Kerri Lyon, in my own car as far as we could permissibly go. We then were taken a little closer to the scene by detectives in an unmarked car, and then eventually Kerri and I started collecting interviews for a piece about the mass civilian exodus, interviewing a number of folks as they steadily streamed out of Manhattan on the Brooklyn Bridge. A number of people were very rattled and the F-16s flying overhead made it even more unsettling. I remember going to area hospitals looking for survivors, but of course they were few and far between. The next day I went out with another reporter, Cheryl Wills, to report on a jittery city. What we came across were a lot of bomb threats and evacuations of transit hubs. The next day I started working on a timeline piece as to what and when things happened, and by week's end I had assembled an hour special that chronicled our reporting on a story that at first had reported some six

215

thousand people dead. The special included a number of press conferences, volunteer efforts, and of course a fire department promotion ceremony in which scores of firefighters were promoted to fill in the depleted ranks of the department's top leadership. So the attempt early on was to try and keep up with a story, the size of which no one in TV news had ever seen.

How did you develop the concept "Post 9/11 A New Reality" series for your coverage on the five-year anniversary? What were the criteria for choosing the subject matter and panelists?

I, along with management colleagues at the station, made an attempt to identify the large stories around the attacks that still continued to resonate nearly five years out. We knew the rebuilding effort was going to be the big story, defined as much by its inaction as much by its slow forward progress. We decided to twin that with memorial plans, as they both were and are still intertwined. The second theme was the health of responders, and what was and wasn't said about air quality in the aftermath of the disaster. The third subject we wanted to touch on was how much in the way of civil liberties were you willing to surrender for added security. Our fourth subject looked at how Hollywood and the lively arts had come to interpret the events of September 11. Our panelists were chosen on the basis of how much fire they could add to the discussion. For instance, our health panel consisted of, among others, a doctor heading up the medical monitoring program at Mt. Sinai, as well as David Worby, an attorney who is representing thousands of rescue and emergency workers suing the city for increased benefits. For the arts discussion, we had John Hoffman, the director of the acclaimed HBO special "In Memoriam," as well Oskar Eustis, the artistic director of the Public Theatre. As in any discussion, you are looking for a mixture of opinions and expertise, and I think we were able to achieve that in our conversations.

How did your plans coincide with content and discussions that took place at the Aftershock conference at large? Was there a par-

*ticular subject related to the attacks and their aftermath that you
believe was particularly important to address at a 9/11 retrospective?*

I think the Aftershock conference may have been a little more
global than our discussions on the subject. I believe we had more
of a mission to try and address concerns that day-in and day-out
registered with residents in the city, like health concerns and
the rebuilding. With a large Muslim community, the questions
arising out of the government's quest for increased scrutiny of
phone calls and internet and library activity is a very current
issue. I think we complemented Aftershock, supporting it without being redundant.

*How is NY1's coverage of the events of 9/11/2001 from a per-
spective of five years later different than the coverage immediately following the attacks? Do you anticipate that your perspective
will continue to change? How so?*

Every story gains perspective within the first few hours of it
being reported. 9/11 is no different. I think as you get further
from this story, you find yourself much more apt to question
decisions that concern efforts to rebuild, that address health
concerns, that attempt to diminish civil liberties in the name of
increased protection, that portray events in the theater or on a
large screen that are a direct result of 9/11. Some of the stories
can be more nuanced, like how do we live our lives now? Do we
wake up in the morning and immediately put on the TV? Do we
question authority more, do we question leadership decisions?
When we see something, do we say something? And perhaps,
has our sense of ourselves as Americans in the world changed?

A CONVERSATION WITH STEVE MENDELSOHN, EXECUTIVE DIRECTOR, PROJECT REBIRTH

Where were you on 9/11/2001?

On September 11, I was on the corner of Wall and Broad Streets when the second plane hit the towers. I heard a loud rumble and felt a wave of heat as if a volcano had exploded. I was then in a shower of debris and paper from the Trade Center, and something scratched my eye requiring a visit to an eye doctor. Fortunately, my eye healed on its own. I had been working for a company located on Broad Street and had to evacuate the staff through the dust cloud. It was all very frightening.

What was it about the events of 9/11 that compelled the Project Rebirth team to come together and formulate an artistic response to the attacks? What were the criteria for choosing the ten people featured in the documentary?

Shortly after 9/11, Jim Whitaker, the founder and director of Project Rebirth, was walking around Lower Manhattan and watching people as they looked at the wreckage of the World Trade Center. While he felt a sense of dread, he was also able to see hope for the future—hope that we as individuals, as a city, and as a nation will somehow get through this. Jim felt the need to create Project Rebirth as a way to document the recovery of the site for generations to come. He wanted to create a film that will serve as a historic legacy telling the story of how we recovered from the attacks of that day. That is why he decided to place time-lapse cameras around the site to capture the complete reconstruction from start to finish. The first cameras started operating in March 2002, the sixth month anniversary of the attacks. Today we have twelve cameras filming the reconstruction from a variety of angles.

219

Jim also wanted to document the human coping and healing processes. That is why we have been filming the lives of ten people every year since 2002. These people were selected because they represent a variety of demographic and personal situations, including a survivor who escaped from the South Tower, a woman who lost her fiancé, a firefighter who escaped from the WTC but lost most of his friends, a young man whose mother died on 9/11, and others.

Part of your job at Project Rebirth has been to relive the tragic events of 9/11 repeatedly over the last five years. Is this something that has been difficult for yourself and the crew?

Actually, as you might imagine, it is sometimes difficult to listen to the multitude of personal stories about 9/11. Before joining Project Rebirth, I thought that I had heard so much about the events that I could handle anything related to the tragedy. What I've realized, though, is that I had suppressed many of my own memories related to that day. Reliving the events reminds me of the feelings I had on September 11 and afterwards. Still, I have learned so much from the subjects of the film. Each one of them had a unique experience and personal trauma. The courage and fortitude that they have shown has impressed me. They are learning to move on in ways that even they thought would be impossible. Watching their progress gives me a lot of hope. It shows me the incredible ability of humans to adapt and grow—no matter what happens. Members of the crew constantly comment on how important the project is for them. They have told us that it is a privilege to work on Project Rebirth and that they are always amazed at how much they have learned from the subjects of the film. Hopefully, Project Rebirth will provide similar inspiration to people all over the world for years to come.

The newly edited version of the film screened at the Aftershock conference presents a renewed sense of hope for the future of Lower Manhattan and the many people whose lives were impacted by 9/11/2001. What is the future of Project Rebirth? Was there a worry at the beginning of the project that the stories and lives depicted would not be as positive or uplifting as they have turned out so far?

Regarding the future of Project Rebirth, we will continue to film the reconstruction at Ground Zero until it is completed. The latest estimate is that all of the buildings will be finished by 2013. We also plan to follow the ten subjects for a total of ten years, through 2011 (the tenth anniversary of the attacks). The finished film will be seen at the Memorial Museum at Ground Zero, and it will also be distributed theatrically around the world. Additionally, all of our original film footage is being donated to the Library of Congress so that it can be accessed by the public for centuries to come.

Project Rebirth was created as a way to capture the recovery processes of the physical site and the people we are following. We had faith that most of the people we selected would learn to move on and rebuild their lives. We understood that some of the stories wouldn't be as positive as the others—that is the nature of life. While the stories have been uplifting, the film does show the struggles that all the subjects have faced. Some have faced more challenges than others. Regardless of the direction the stories follow going forward, we believe that the film will show a variety of responses to tragedy which will help people in the future as they pursue their own path of recovery from different situations, whether it be natural disasters like Hurricane Katrina, man-made events like 9/11, or personal and family difficulties.

What did the Aftershock conference, at large, make you think about as you continue to pursue this very important professional and artistic endeavor? The themes and subject matter of your film brilliantly and poignantly reflect many of the topics that were discussed at the conference at large. Was there anything particularly unique that you took away from Aftershock that inspired you to continue your work, seeing as you are now at the halfway point of your endeavor?

221

The Aftershock conference provided a comprehensive overview of the different aspects of life since 9/11/01. I was impressed by the variety of viewpoints that were expressed throughout the conference. One thing that was made clear at the conference is that regardless of whether a person agrees with the path pursued by private and governmental leaders, there is no doubt that the City and the nation have exhibited phenomenal resiliency. New York has never been as strong as it is today. Lower Manhattan has been completely transformed into a vibrant, multi-purpose community—something that seemed unfathomable at the end of 2001. Similarly, the progress shown by the subjects of our film has exceeded our expectations. Yes, we knew they would learn to move on. But like the City, many of them have dramatically reshaped their lives and have become stronger than before. It is that realization that inspires us to continue our work.

BIOGRAPHIES

KEYNOTE SPEAKERS

DAVID GERGEN is a commentator, editor, teacher, public servant, best-selling author and adviser to presidents—for 30 years, David Gergen has been an active participant in American national life. He served as director of communications for President Reagan and held positions in the administrations of Presidents Nixon and Ford. In 1993, he put his country before politics when he agreed to first serve as counselor to President Clinton on both foreign policy and domestic affairs, then as special international adviser to the president and to Secretary of State Warren Christopher.

David Gergen currently serves as editor-at-large at *U.S. News & World Report*. He is a professor of public service and the director of the Center for Public Leadership at the John F. Kennedy School of Government. Mr. Gergen also regularly serves as an analyst on various news shows, and he is a frequent lecturer at venues around the world. He served as a moderator of World @ Large, the 13-part PBS discussion series, for the past two seasons. In the fall of 2000 he published a book titled, *Eyewitness to Power: The Essence of Leadership, Nixon to Clinton.*

In the past, Mr. Gergen has served in the White House as an adviser to four Presidents: Nixon, Ford, Reagan, and Clinton. Most recently, he served for eighteen months in the Clinton administration, first as Counselor to the President and then as Special Advisor to the President and the Secretary of State. He returned to private life in January 1995.

From 1984 to 1993, Mr. Gergen worked mostly as a journalist. For some two-and-a-half years, he was editor of *U.S. News*. Working with the owner and editor-in-chief, Mortimer Zuckerman and a revived staff, he helped to guide the magazine to record gains in circulation and advertising. During that period, he also teamed up with Mark Shields for political commentary every Friday night for five years on the MacNeil/Lehrer

NewsHour. The two were a popular political team and won numerous accolades for their political coverage.

A native of Durham, North Carolina, Mr. Gergen is an honors graduate of Yale University (A.B., 1963) and the Harvard Law School (LL.B., 1967). He is a member of the D.C. bar. In addition, Mr. Gergen served for three-and-a-half years in the U.S. Navy, where he was posted for about two years to a ship home-ported in Japan. Mr. Gergen is active on many non-profit boards and is Chairman of the National Selection Committee for the Ford Foundation's program on Innovations in American Government. He frequently lectures here in the United States and overseas and holds fourteen honorary degrees.

Mr. Gergen has been married since 1967 to Anne Gergen of England. She is a family therapist and they live in Cambridge, Mass. They have two children, Christopher and Katherine.

DORIS KEARNS GOODWIN was born in Brooklyn, New York and grew up in Rockville Center, Long Island. Her invalid mother encouraged her love of books, while her father shared her love of baseball; she traces her interest in history to her childhood experience recording the fortunes of the Brooklyn Dodgers.

She received her B.A. from Colby College, Maine, graduating Magna Cum Laude. While in college, she undertook summer internships at the U.S. Congress and the State Department. She won a Woodrow Wilson Fellowship and earned a Ph.D. in Government at Harvard University.

She was serving as a White House Fellow in 1967, when her opposition to President Johnson's foreign policy led her to co-author an article for *The New Republic* entitled "How to Remove LBJ in 1968." Only a few months later, she became a special assistant to President Johnson in the White House. The President apparently believed that having a White House fellow who was critical of the administration would prove he did not feel threatened by the growing anti-war sentiment in America

After President Johnson's retirement in 1969, Doris Kearns began a decade's work as a Professor of Government at Harvard, where she taught a course on the American Presidency. On weekends, holidays and vacations she traveled

224

to Johnson's ranch in Texas, to assist the ex-president in the preparation of his memoir, *The Vantage Point* (1971).

President Johnson died in January, 1973. In 1975, Doris Kearns married Richard Goodwin, who had been an advisor and speechwriter to Presidents Kennedy and Johnson and to Sen. Robert Kennedy. In 1977, Doris Kearns Goodwin published her first book, *Lyndon Johnson & the American Dream*, drawing on her own conversations with the late president. It became a *New York Times* bestseller and Book of The Month Club selection. With her husband's assistance, she began research in the Kennedy family archives in Hyannisport. The result was The *Fitzgeralds & The Kennedys* (1987), a *New York Times* bestseller for five months. In 1990, it was made into a six hour miniseries for ABC Television.

Her next success was *No Ordinary Time: Franklin and Eleanor Roosevelt: The American Homefront During World War II* which was awarded the Pulitzer Prize for history in 1995. *Wait Till Next Year: A Memoir* was published in 1997. Her tale of growing up in the 1950's and her love of the Brooklyn Dodgers became a *New York Times* bestseller and Book of the Month Club selection.

In addition to her books, Ms. Goodwin has written numerous articles on politics and baseball for leading national publications. She is a regular panelist on Public Television's *The News Hour with Jim Lehrer* and a frequent commentator on NBC and MSNBC. She has been consultant and on-air person for PBS documentaries on LBJ, the Kennedy family, Franklin Roosevelt and Ken Burns's *History of Baseball*. She is also the first woman ever to enter the Red Sox locker room. Doris and Richard Goodwin have three sons. They make their home in Concord, Massachusetts.

LEE HAMILTON is a former U.S. Congressman, vice chair, National Commission on Terrorist Attacks upon the United States, and president and director of the Woodrow Wilson International Center for Scholars. Prior to becoming director of the Woodrow Wilson Center in 1999, Hamilton served for thirty-four years in Congress representing Indiana's Ninth District.

During his tenure, he served as chairman and ranking member of the House Committee on Foreign Affairs (now the Committee on International Relations), chaired the Subcommittee on Europe and the Middle East from the early 1970s until 1993, the Permanent Select Committee on Intelligence, and the Select Committee to Investigate Covert Arms Transactions with Iran. Hamilton also served as chair of the Joint Economic Committee, working to promote long-term economic growth and development. As chairman of the Joint Committee on the Organization of Congress and a member of the House Standards of Official Conduct Committee, he was a primary draftsman of several House ethics reforms. Since leaving the House, Mr. Hamilton has served as a commissioner on the influential United States Commission on National Security in the 21st Century (the Hart-Rudman Commission), and was co-chair with former Senator Howard Baker of the Baker-Hamilton Commission to Investigate Certain Security Issues at Los Alamos. He is currently a member of the President's Homeland Security Advisory Council. Mr. Hamilton is a graduate of DePauw University and Indiana University Law School, as well as the recipient of numerous honorary degrees and national awards for public service. Before his election to Congress, he practiced law in Chicago and Columbus, Indiana.

WILLIAM KRISTOL is renowned as a perceptive strategist and analyst of American politics, William Kristol unravels the intricacies of Washington with intelligence and insight.

Perhaps no one has been more influential in setting the current political agenda than William Kristol. He is founder and editor of the prominent Washington-based political magazine *The Weekly Standard*, a magazine that's often considered a "must read" for anyone who wants to understand American politics and society as a whole.

Unafraid to criticize even those on his side of the political fence, Kristol draws on all aspects of his background, domestic and foreign policy, and the daily headlines to explain American politics today. His unparalleled insight provides audiences with an overview of the issues shaping election strategies, policies and tactics. As a regular on FOX News Sunday, Kristol has spearheaded the foreign policy debate on the war in Iraq and has played a central role in virtually every major political drama of the last decade.

Kristol, who teaches political science at Harvard's Kennedy School of Government, offers a behind-the-scenes perspective of someone who has participated in national politics at the highest levels. Author of *War Over Iraq* and editor of *The Weekly Standard: A Reader 1995-2005*, he helps promote global leadership as a co-founder of the Project for the New American Century, whose goal is to promote American global leadership.

DANIEL L. DOCTOROFF is Deputy Mayor for Economic Development and Rebuilding for the City of New York. Under the leadership of Mayor Michael R. Bloomberg, Mr. Doctoroff has overseen one of the city's most dramatic economic resurgences, spearheading the effort to reverse New York's fiscal crisis after the attacks of 9/11 through a five-borough economic development strategy. This plan includes the most ambitious land-use transformation in the city's modern history; the largest affordable housing program ever launched by an American city; the formation of new Central Business Districts and Industrial Business Zones; and the creation of new, great destinations like the Harbor District, which will link together new parkland and miles of waterfront esplanades in Lower Manhattan, Governors Island, and Brooklyn. By focusing on making New York's economy more diverse, its business climate more hospitable, and its communities more livable, Mr. Doctoroff has helped lead New York to its strongest economic position in decades. In 2005, the city achieved record levels of jobs, visitors, population, and the greatest number of housing starts since the 1960s.

Prior to joining the Bloomberg administration, Mr. Doctoroff was Managing Partner of Oak Hill Capital Partners, one of the oldest and most respected private equity investment firms. During his fourteen-year association with Oak Hill, Mr. Doctoroff led the purchases of companies in a wide variety of industries, including information services, insurance, thrifts, cable television, hotels and leasing. While at Oak Hill, Mr.

227

Doctoroff founded NYC2012, the organization dedicated to bringing the Olympic Games to New York. He continued to oversee the New York's bid as Deputy Mayor, ensuring that the Olympic effort spurred parks, housing, and economic development projects in all five boroughs.

Prior to joining Oak Hill, Mr. Doctoroff was an investment banker at Lehman Brothers. Mr. Doctoroff received a B.A. degree from Harvard College. He received a J.D. degree from The Law School at the University of Chicago. Before attending law school, Mr. Doctoroff was a political pollster.

PANELISTS

JOE BACZKO is the Dean of the Lubin School of Business at Pace University. Mr. Baczko (pronounced "BASS-co"), a Hungarian who was born in Germany, raised in France, and is a U.S. citizen, earned his B.S. from the School of Foreign Service at Georgetown and was assistant director of admissions there before serving as an officer in the U.S. Marine Corps in Vietnam. He then earned his MBA from the Harvard Business School. In a career of guiding organizations to market share leadership and significant profitability, he has been Chairman and CEO of Frank's Nursery and Crafts, Inc.; President and COO of Blockbuster Entertainment Corporation during a period when it doubled its size and operated more than 3000 stores worldwide; Founder and President of the International Division of Toys R Us, Inc., for which he opened subsidiaries and joint ventures in eleven countries; and CEO of Max Factor (Europe). He currently lives in New York City, and is a consultant to private equity firms on acquisitions in consumer products, specialty retailing, and consumer services sectors.

SADIE C. BRAGG is Senior Vice President of Academic Affairs and Professor of mathematics at Borough of Manhattan Community College, CUNY. Dr. Bragg served as American Mathematical Association of Two-Year Colleges (AMATYC) president from 1997-1999 and is currently a co-director of

Project ACCCESS (Advancing Community College Careers: Education, Scholarship and Service), an AMATYC professional development program. Dr. Bragg holds a doctorate in the College Teaching of Mathematics from Teachers College, Columbia University.

JOHN CAHILL was Secretary and Chief of Staff to former New York Governor George Pataki. He was first named to the Governor's staff in 2001 as a Senior Policy Advisor. Before his appointment in the Governor's office, he served as Commissioner of the New York State Department of Environmental Conservation (DEC). Prior to that, Cahill worked in a private law practice for ten years. He holds a bachelor's degree in Economics from Fordham University. He earned his law degree, as well as a Master's degree in Environmental Laws, from Pace University.

DAVID A. CAPUTO became the sixth president of Pace University in July 2000 and is presiding over its 100th anniversary in 2006. He leads a private, comprehensive metropolitan university with 14,000 students, seven campuses, and a growing national reputation for offering students opportunity, teaching, and learning based on research, civic involvement, international perspectives, and measurable outcomes.

During Dr. Caputo's tenure at Pace he has developed a five-year strategic plan and accelerated its regional and national leadership. To make the cost of education predictable, Pace was one of the first universities to guarantee flat tuition for up to five years and graduation in four. The Pace Academy for the Environment has forged a consortium of 32 colleges and universities in the Hudson River Valley to address environmental concerns. A Center for Downtown New York, established after the attacks of 9/11, and the Pace Downtown Index of Economic Activity have involved the university in the revitalization of Lower Manhattan; the university's Pace Poll has advanced understanding of civic issues.

Before his appointment at Pace, Dr. Caputo served for five years as president of Hunter College. Prior to that he was at

Purdue University for 26 years, last serving as Dean of its School of Liberal Arts.

A political scientist, Dr. Caputo earned his B.A. in Government from Miami University in Oxford, Ohio, and his M.A. and Ph.D. in Political Science from Yale University. His research and teaching interests are urban and intergovernmental relations, election reporting and administration, and Italian politics. A member of Phi Beta Kappa, he has been the recipient of a Senior Fulbright Chair appointment at the University of Bologna, National Science Foundation Faculty Fellowship, a Lilly Endowment Fellowship, and a Visiting Fellowship at Princeton University's Woodrow Wilson School of Public and International Affairs. He has also served as a Visiting Scholar at Harvard University's Center for Population Studies.

On the national level, Dr. Caputo serves as a director of the National Association of Independent Colleges and Universities, on the Council of Presidents of the Association of Governing Boards of Universities and Colleges, and on the American Council on Education's Commission for Lifelong Learning. He chairs the New York State Conference of Independent Colleges and Universities and is a member of the New York City Mayor's Election Modernization Task Force, a director of the Lower Manhattan Cultural Council and the Westchester Arts Council, and a board member of the Westchester County Association and chair of its Education Committee.

JAMES CAVANAUGH is the President and Chief Executive Officer of the Hugh L. Carey Battery Park City Authority, a public benefit corporation created by the New York State Legislature to develop and maintain a 92-acre parcel of land located on the southern tip of Manhattan across from the World Trade Center site. Under Mr. Cavanaugh's leadership the Battery Park City Authority continues to undertake development of its remaining residential and commercial sites, with completion of the master plan expected to take place by 2009. The Authority is the leader in sustainable building design, having recently enhanced its standards for energy use reduction, recycling, green power generation, and other elements of green. Prior to being named

President & CEO, Mr. Cavanaugh served as the Authority's Chief Operating Officer.

Before joining the Authority, Mr. Cavanaugh was elected to the post of Supervisor for the Town of Eastchester, a community of 32,000 people located in Westchester County, New York. He served in that position for ten years and was best known for his work rebuilding the public infrastructure and controlling taxes. Prior to being elected Supervisor, Mr. Cavanaugh served as a local Councilman.

Mr. Cavanaugh remains a member of the board of the Eastchester School Foundation, and is a member of the Strategic Planning Committee for Lawrence Hospital in Bronxville, NY.

He resides in Westchester County with his wife and two children.

STEVEN CLEMONS is a centrist American blogger and the publisher of the popular political blog *The Washington Note*

Clemons is the executive vice president of the New America Foundation, and the director of the Japan Policy Research Institute. Clemons is also the former executive vice president of the Economic Strategy Institute, former executive director of the Nixon Center for Peace and Freedom, and served as Senator Jeff Bingaman's Senior Policy Advisor on Economic and International Affairs. He has also served on the advisory board to the Center for U.S.-Japan Relations at the Rand Corporation. Earlier in his career, Clemons was the executive director of the Japan America Society of Southern California from 1987-1994.

Clemons also serves on the Board of Advisors of the C.V. Starr Center for the Study of the American Experience at Washington College in Chestertown, Maryland, and the Clarke Center at Dickinson College in Carlisle, Pennsylvania.

JOHN CRONIN began his career as a commercial fisherman before becoming the nation's first full-time Riverkeeper in 1983, responsible for bringing to justice polluters on the Hudson River. In 1999 he brought his drive and commitment to Pace, as a resident scholar in environmental studies and founding member of the Pace Environmental Litigation Clinic and the Institute for Environmental and Regional Studies, both working to protect the Hudson. He kept his crusade active in the classroom, helping students in his "Issues in Politics" course research, design, and lobby for environmental legislation. The result of their efforts, the Hudson River Marine Sanitation Act, was signed into law by Governor George Pataki in 1999. Over the years, Cronin and students from the Environmental Litigation Clinic at the School of Law have brought over 150 legal actions against polluters.

Most recently, the author and former commercial fisherman was named Director of the Pace Academy for the Environment (PAE). The PAE seeks to integrate the University's resources with those of the surrounding community in order to create policies, practices and ideas that sustain the mutually beneficial relationship between nature and society.

ERIC DEUTSCH is the President of the Alliance for Downtown New York, Inc. The mission of the Downtown Alliance is to provide Lower Manhattan's historic financial district with a premier physical and economic environment, advocate for businesses and property owners, and promote the area as a world-class destination for companies, workers, residents, and visitors. The Downtown Alliance manages the Downtown-Lower Manhattan Business Improvement District, serving an area roughly from City Hall to the Battery, from the East River to West Street.

Mr. Deutsch is a graduate of George Washington University, with a B.A. in Political Science, and attended the Columbia University Graduate School of Architecture, Planning & Preservation, where he is a candidate for a MS in Real Estate Development. He currently serves on the Boards of the Battery Conservancy, the Society for the Reservation of Weeksville, and the Brooklyn Navy Yard.

He lives in Park Slope, Brooklyn, with his wife and two children.

Biographies

DR. MICHAEL DOLFMAN currently serves as Regional Commissioner for the United States Bureau of Labor Statistics in New York City. In this position, he supervises and directs all economic analysis and information services for the Bureau throughout the nation, and is responsible for activities in regional offices located in Boston, Philadelphia, Atlanta, Dallas, Kansas City, Chicago, and San Francisco, as well as in New York.

Dr. Dolfman received his undergraduate degree from Albright College; a Master's degree from the University of North Carolina; and a Ph.D. from the University of Pennsylvania. He has published numerous articles in academic journals. Two, written for the Bureau of Labor Statistics, have achieved national and international recognition: "9/11 and the New York City Economy: A Borough-by-Borough Analysis" and "100 Years of Consumer Spending: A Comparison of the Nation, New York, and Boston."

JIM DWYER was a member of the team at *New York Newsday* that won the Pulitzer Prize for sports reporting in 1992. In 1995, again as a columnist with *Newsday*, he received the Pulitzer Prize for Commentary. At present, he is a reporter with the *New York Times*.

Dwyer is the author or co-author of four books. His latest, *102 Minutes: The Untold Story of the Fight to Survive Inside the Twin Towers*, co-written with Kevin Flynn, was a 2005 National Book Award finalist. With other reporters at the *Times*, Dwyer conducted an intensive investigation of what happened inside the Twin Towers at the World Trade Center before they collapsed. The book documents extraordinary but little-known rescues, including the work of Pablo Ortiz and Frank DeMartini, who rescued scores of people from behind jammed doors on the upper floors of the north tower.

A native New Yorker, Dwyer wrote columns for *New York Newsday* and the *New York Daily News* before joining the *Times*. He earned a bachelor's degree in general science from Fordham University in 1979 and a Masters degree in journalism from Columbia University in 1980.

MICHAEL EMMERMAN graduated from C.W. Post with a concentration in international finance. Mr. Emmerman serves as director of the Special Operations Group, a not-for-profit organization that provides research and technical advice to law enforcement and public safety agencies. He has assisted at many disaster scenes including the TWA Flight 800 recovery operation and at the World Trade Center on September 11, 2001. He is also a trustee at Long Island University.

PROFESSOR ED GALEA is the founding director of the Fire Safety Engineering Group (FSEG) at the University of Greenwich, where he has worked in fire safety research since 1986. His work began after the tragic Manchester Boeing 737 fire, when he was commissioned by the U.K. Civil Aviation Authority to simulate the spread of fire and smoke in the disaster. Since then his research has expanded to include the modelling of evacuation, people movement, fire/smoke spread, combustion and fire suppression in the built environment, rail, marine and aviation environments. Professor Galea is the author of over 100 academic and professional publications related to fire. He serves on a number of national and international standards and safety committees concerned with fire and evacuation. His research and consultancy activities have been supported by a wide range of European and North American organizations.

ALICE GREENWALD is the Executive Vice President for Programs, and Director of the Memorial Museum for the World Trade Center Foundation. Prior to that, she was the Director of Programs for the United States Holocaust Memorial Museum.

From 1986-2001, Greenwald was the principal of Alice M. Greenwald Museum Services, providing consulting expertise to a variety of clients. She served as guest curator for the Historical Society of Princeton's exhibition, "Old Traditions, New Beginnings: 250 Years of Princeton Jewish History" and authored the historical essay in the companion catalogue publication published in 2002. Her other clients included the Baltimore Museum of Industry, the Pew Charitable Trusts, and the National Museum of American Jewish History.

Prior to the establishment of her consulting business, Greenwald served as Executive Director of the National Museum of American Jewish History in Philadelphia, Acting Director of the Hebrew Union College Skirball Museum, and Curatorial Assistant at the Maurice Spertus Museum of Judaica, Chicago.

Ms. Greenwald has served as Chairman of the Council of American Jewish Museums and was the recipient of several awards including a Fellowship for Museum Professionals (National Endowment for the Arts) and a Danforth Graduate Fellowship. She is a graduate of Sarah Lawrence College and holds an A.M. in the History of Religions from the University of Chicago Divinity School.

ROBERT HACKMAN is a Graduate Student at Pace University's Seidenberg School of Information Systems. He is a native and citizen of the Republic of Ghana. Mr. Hackman is currently enrolled in the Master of Science program in Internet Technology.

TOM HEALY is president of the Lower Manhattan Cultural Council. Founded by David Rockefeller in 1973, LMCC was located in the World Trade Center until September 11, 2001. LMCC's offices, artist residency, and performance venues were destroyed in the attack, and tragically an LMCC artist died that day. Since 2004, Healy has led LMCC's comeback and rapid transformation into one of the city's leading arts organizations.

Healy founded one of the first contemporary art galleries in Chelsea in 1994, and represented several emerging artists who have gone on to prominent careers, including Tom Sachs, Venice Biennale winner Janet Cardiff, and performance artist Karen Finley. He has served on the boards of Creative Time, PEN America, Poet's House, the List Center for Art and Politics, the Architecture Committee of the Whitney Museum, FENCE magazine, ArtOmi, and the Civic Alliance. His poetry and essays have been published widely. He is currently producing the first independent feature film of the Neistat Brothers.

KENNETH T. JACKSON is a professor of History and Social Sciences at Columbia University. A frequent television guest, he is best known as an urban historian and a preeminent authority on New York City, where he lives on the Upper West Side.

Jackson was born in Memphis, Tennessee, earning his B.A . in 1961 at the University of Memphis and his Ph.D. in 1966 at the University of Chicago. He served as an assistant professor for the Air Force Institute of Technology at Wright-Patterson Air Force Base from 1965-1968 and then joined the Columbia faculty as an assistant professor in 1968, earning his tenure by 1970.

Jackson's achievements as an author include *The Klu Klux Klan in the City, 1915-1939* (1961), *Cities in American History* (1972), *Crabgrass Frontier: The Suburbanization of the United States* (1985), and *The Encyclopedia of New York City* (1995) for which he served as the primary editor.

Jackson has earned numerous distinctions as a professor at Columbia University where he is the director of the Herbert H. Lehman Center for American History and the Jacques Barzun Professor of History and Social Sciences. Jackson teaches a popular lecture at the university on "The History of the City of New York." Jackson has also served as president of the Urban History Association, the Society of American Historians, the Organization of American Historians, and the New York Historical Society.

BEVERLY KAHN is Associate Provost for Academic Affairs. She came to Pace in 2001 from Fairfield University where she was Dean of Arts and Sciences. Prior to that, she held faculty positions in Political Science at Ohio State University and the University of South Carolina.

Dr. Kahn received her A.B. in Political Science from Dickinson College, her M.S. Columbia University, and her Ph.D. from Indiana University. Her research focuses on Italian politics and Italian political philosophy. She was the recipient of a Fulbright Research Scholarship for Italy as well as the Rome Prize from the American Academy in Rome. She is also the recipient of a Fulbright travel award to Japan and South Korea.

CHARLES LAI is the founder of the Museum of Chinese in the Americas (MoCA), and came to New York from China in 1965. Lai, his five siblings, and their parents moved into a one-bedroom apartment in a tenement building on the outskirts of Chinatown. His father took a job as a cook in a restaurant and his mother found work in a garment factory.

Lai has a distinguished record of public service, including serving as director of programs and planning at the Asian American Federation of New York, where he was responsible for implementing the Federation's "9/11 Relief, Recovery, and Rebuilding Initiative." Prior to that, he served as the executive director of the Chinatown Manpower Project, a vocational training program, and the Director of Policy and Budget for former Manhattan Borough President, Ruth Messinger.

After graduating from Princeton University, Lai returned to his neighborhood armed with a good education and a passion to help his fellow Chinese Americans, who had expanded Chinatown from an eight-block neighborhood into an expansive and ever-growing community. The question of what defined the community posed a challenge for him.

"In the spectrum of those who have been here for generations or for those who are new, we each had a different view," Lai says. "Out of that, was there something that would bind us?"

By the mid-1970s, Lai found others who were also grappling to define their cultural and communal identity. In 1980, he met Jack Tchen, with whom he started a community-based organization called the New York Chinatown History Project. This early organization eventually grew into today's MoCA.

DR. BRUCE LOGAN is the President and CEO of the New York Downtown Hospital, where he has worked for 25 years. His specialty is in Internal Medicine.

A native of Salem, Massachusetts, Logan has worked in New York since he came to attend Columbia University Medical School in 1967. He and his wife life in South Orange, N.J. They have two daughters.

JULIE MENIN is the President and Founder of Wall Street Rising, a not-for-profit organization dedicated to revitalizing Lower Manhattan. She is also Chairperson of Community Board One and serves on four governmental committees dedicated to the development of Lower Manhattan, including two advisory councils of the Lower Manhattan Development Corporation. Previously, she was Senior Regulatory Counsel for Colgate-Palmolive. Menin is a magna cum laude graduate of Columbia University and the Northwestern University School of Law. She lives in Manhattan.

JOHN MERROW began his career as an education reporter with National Public Radio in 1974 when he created "Options in Education." That weekly series received more than two dozen broadcasting awards, including the George Polk Award in 1982. His first television series, "Your Children, Our Children" earned an Emmy nomination for Community Service in 1984.

From 1985 to 1990, Merrow was the education correspondent for *The MacNeil/Lehrer News Hour.* In 1993 he created *The Merrow Report* for PBS, followed by the NPR series of the same name in 1997. In 2000 he returned to the NewsHour to provide reports on education, and in 2002 he and his colleagues began producing programs for the PBS series *Frontline.*

Merrow has received the George Foster Peabody Award, the George Polk Award, the Hugo Award from the Chicago International Film Festival, two CINE Golden Eagles, eleven consecutive awards from the Education Writers Association, and, in 2005, another Emmy nomination.

Merrow received an AB from Dartmouth College in 1964, an MA in American Studies from Indiana University in 1968, and a doctorate in Education and Social Policy from the Harvard Graduate School of Education in 1973.

CONGRESSMAN JERROLD NADLER represents New York's Eighth Congressional district. The Eighth, one of the most diverse districts in the nation, includes Manhattan's West Side below 89th Street, Lower Manhattan, and areas of Brooklyn including Borough Park, Coney Island, Brighton Beach, Sea Gate, Bay Ridge, and Bensonhurst.

Congressman Nadler was first elected to the House of Representatives in 1992 after serving for sixteen years in the New York State Assembly. Throughout his career he has championed civil rights, civil liberties, efficient transportation, and a host of progressive issues such as access to health care, support for the arts and protection of the Social Security system. He is considered an unapologetic defender of those who might otherwise be forgotten by American law or the economy, and is respected specifically for his creative and pragmatic legislative approaches.

From his leadership in response to the September 11 terrorist attacks on his district, to his insight and policymaking prominence on issues facing Israel and the Middle East, Congressman Nadler has constantly sought to be steadfast and responsive in his service to New York and the nation.

DAVID M. NEWMAN is an industrial hygienist on the staff of the New York Committee for Occupational Safety and Health (NYCOSH), a private, non-profit, union-based membership organization. Since September 11, 2001, he has coordinated NYCOSH's World Trade Center Health and Safety Project, which, in partnership with the National Disaster Ministries of the United Church of Christ, has provided education and technical assistance on 9/11-related occupational and environmental health issues to workers, unions, employers, and community and tenant organizations throughout Lower Manhattan.

JOESPH PETRO is the Executive Vice President and Managing Director of Citigroup Security and Investigative Services. He is responsible for fraud prevention and investigations, physical security, and executive protection for Citigroup and its worldwide subsidiary companies. In 1993, he joined Travelers Group, Citigroup's legacy company, as Director of Corporate Security.

From 1971 to 1993, Petro was a special agent and senior executive with the United States Secret Service where he served in numerous operational and management positions, including supervising the Presidential and Vice Presidential protective divisions and the Washington Field Office. Prior to his career in the Secret Service, Petro was a Lieutenant in the U.S. Navy's River Patrol Forces.

Currently, Petro is the Co-Chairman of the State Department's Overseas Security Advisory Council, a member of the Board of Directors of the International Security Management Association, and Vice-Chairman and member of the Board of the New York City Law Enforcement Explorers Council.

Petro is a graduate of Temple University and was a Fellow at Princeton University's Woodrow Wilson School of Public and International Affairs. He is the author of *Standing Next to History, An Agent's Life Inside the Secret Service*.

STEFAN PRYOR is the Deputy Mayor of Newark, New Jersey. Prior to that he served as President of the Lower Manhattan Development Corporation. Appointed to the top spot in 2005, he had previously served the agency as Senior Vice President for Policy and Programs. Pryor has a long track record in downtown New York—he was a Vice President at the Partnership for New York City and helped to coordinate downtown business reinvestment after 9/11. He was also the LMDC's first official employee.

SALLY REGENHARD is the founder and chairperson of the Skyscraper Safety Campaign. The Skyscraper Safety Campaign was created in December 2001 by the Regenhard Family, in memory of Ms. Regenhard's beloved son, Christian Michael Otto Regenhard, a 28-year-old Probationary Firefighter who remains missing at the WTC, along with his entire Engine Company 279, to this date. The Skyscraper Safety Campaign was created in part to foster increased oversight of the construction and maintenance of building with regards to safety and security. Ms. Regenhard testified in November 2003 before the 9/11 Commission and has remained a powerful advocate for the victims and the families of September 11.

THOMAS H. ROGER is a founding member of Families of September 11, and serves on its board. He has been a committed advocate for 9/11 families, raising awareness and personally working on such issues as airline security, the Victims Compensation Fund, 9/11 curriculum in schools, and the memorial at Ground Zero. Roger has been prominently involved in the memorial process serving as a member of the Lower Manhattan Development Corporation's Families Advisory Council, the Memorial and Museum Program Committees, and the World Trade Center Memorial Foundation Board of Directors. His twenty-four-year-old daughter, Jean Roger, was a flight attendant on American Airlines Flight 11.

WILLIAM RUDIN was encouraged by his grandfather, Samuel Rudin, to join the family real estate business in 1979. He has put his own personal mark on the business by championing the resurgence of Lower Manhattan by giving new life to old properties, both residential and commercial, converting them into technology smart buildings. In 1995, after remaining vacant for five years, 55 Broad Street—*www.55broadst.com*—became the first building to bring broadband connectivity to each and every floor. With this innovative concept, 55 Broad Street was leased in 18 months and led the turnaround in downtown Manhattan while starting the trend to make smart, technology ready buildings. Other projects include Wired@110 Wall Street, 3 Times Square – The Reuters Building, and Rudin's latest project, the Global Connectivity Center at 32 Sixth Avenue.

Rudin serves as the Chairman of the Board for the Battery Conservancy; as a member of the Boards of the Metropolitan Museum of Art and New York University; Vice Chairman of the Real Estate Board of New York; and an Executive Committee Member of the New York Hall of Science.

Rudin graduated from New York University's School of Business and Public Administration in 1979 with a Bachelor of Science and currently lectures at various schools.

JOSEPH RYAN is the Criminal Justice & Sociology Chairperson at Pace University. He is also a twenty-five year veteran of the New York City Police Department. Other credentials include: Former Visiting Fellow, National Institute of Justice; former member, NYC Mayor's Task Force on Child Abuse and Neglect; former member, Center for Disease Control's Violence Prevention Initiative; member, American Society of Criminology.

MARK SCHAMING is the Director of Exhibitions at the New York State Museum. Schaming is responsible for the planning, design, production, and installation of 125,000 square feet of gallery space.

Schaming designed "The World Trade Center: Rescue Recovery Response" exhibit at the NYSM, which was the first permanent exhibition of artifacts documenting the September 11 attack. He spent over 40 days at the Fresh Kills Landfill working with the NYPD, NYFD, FBI, and other key recovery effort personnel to facilitate the collection, documentation, and preservation of World Trade Center artifacts.

Schaming designed the September 11 timeline installed at the WTC site with the Families of September 11 and the Port Authority of New York and New Jersey. Working now with a number of September 11 related projects, Schaming was a member of the WTC Memorial Center Advisory Committee, works as a consultant to the Tribute Center at the World Trade Center site, the FDNY, PANYNJ, and various other city and state agencies. He received a Master's in Fine Arts from the University of Albany and a Bachelor's in Fine Arts and Art History from the State University of New York at Buffalo.

DR. RICHARD SHADICK is director of Pace University's Counseling Center . He is also director of the Trauma Response Service of the White Institute and a member of the New York State Disaster Response Network. He frequently presents at conferences on issues related to 9/11, suicide, trauma, and complicated bereavement; as well, he actively engages in editorial work on a number of psychology journals, and has a private practice in lower Manhattan.

Shadick was involved in a study developing a new treatment to counsel those who are grieving the loss of a loved one from 9/11; was a lead trainer for the Mental Health Association's grant to train 5,000 mental health professionals to respond to future terrorist attacks; has counseled Ground Zero workers; and worked clinically with survivors of 9/11.

STEVEN SPINOLA is President of the Real Estate Board, the real estate industry's leading trade association in New York City. The Real Estate Board's membership includes over 6000 building owners, developers, brokers, managers, banks, insurance companies, brokerage houses, architects, attorneys, and other individuals and institutions professionally involved in New York City real property. The Board is a vigorous advocate of policies to promote local economic growth and residential construction and rehabilitation. Crain's *New York Business* has included Mr. Spinola in its list of the 100 Most Influential New Yorkers, describing him as the "authoritative public voice, lobbyist, and conscience" of the industry he represents.

Mr. Spinola holds a Bachelor of Arts degree from the City College of New York with a concentration in political science and government. He attended the Harvard Business School/Kennedy School of Government Summer Program for Senior Managers in Government.

NIKKI STERN has served as a facilitator and advisor to countless public processes associated with the memorializing and rebuilding efforts at Ground Zero. Ms. Stern is a member of the Lower Manhattan Development Corporation's Families' Advisory Council; wrote the original iteration of the WTC Memorial Mission Statement; and served on the committee that drafted the version that was used as part of the final Memorial Design competition. Stern also served as the Director of Families of September 11, a national families' advocacy group, as well as the New Jersey Governor's 9/11 Victims' Families Liaison.

Nikki Stern holds a B.A. in History from Washington University and an M.A. in Political Science from Georgetown University. Her husband, James Potorti, a Vice President with Marsh and McLennan, was lost at the World Trade Center on September 11, 2001.

EUGENE STEUERLE is a founder of Our Voices Together, a non-profit, nonpartisan organization started by September 11 families and others to help improve the lives of families worldwide and build international understanding. His wife, Norma, whom he met as a student at the University of Dayton, died in the attack on the Pentagon. He is a senior fellow at the Urban Institute, the author or co-author of eleven books, and a former deputy assistant secretary of the U.S. Department of the Treasury and former president of the National Tax Association. A 1968 mathematics graduate, Dr. Steuerle was the original organizer and coordinator of the U.S. Department of Treasury study that led to the Tax Reform Act of 1986, perhaps the most important reform of the U.S. income tax code. In 2004, he received the Distinguished Alumnus Award from the University of Dayton.

LORNA THORPE is the Deputy Commissioner for the New York City Department of Mental Health and Hygiene, Division of Epidemiology. Her work as a leading epidemiologist for the Centers for Disease Control sent her around the globe tracking the spread of infectious diseases. After the terrorist attacks against the United States on September 11, Thorpe was assigned, along with other leading epidemiologists, to work in New York City to assess the lingering affects of the attack. She was the recipient of the 2002 Paul C. Schnitker International Health Award, which was presented by the Epidemic Intelligence Service.

DAVID WARREN is currently President of the National Association of Independent Colleges and Universities (NAICU). He came to NAICU in 1993 after serving nearly a decade as president of Ohio Wesleyan University.

Warren has been a tireless crusader for America's private colleges and for increased financial aid funding for all college students. Dr. Warren is widely considered one of the most persuasive and influential voices for higher education within Washington D.C. Beyond NAICU, he has orchestrated and led cooperative efforts with the other major higher education associations. He has co-chaired the National Campus Voter Registration Project which, in each presidential and congressional election since 1998, has engaged the nation's campuses in the political and electoral process. He also has spearheaded the Student Aid Alliance, an ongoing campaign to expand student aid which has resulted in a 62 percent increase in the Pell Grant, as well as CampusCares, an initiative to gain national recognition for the community service and civic engagement contributions by America's colleges and universities. Currently, Dr. Warren is co-chairing Keep College Possible: The Campaign for the Reauthorization of the Higher Education Act.

The recipient of 12 honorary degrees and numerous other academic and civic awards, Warren has been described by former American Council on Education President Stanley O. Ikenberry as one of the most gifted leaders and creative thinkers in all of American higher education.

KATHRYN WYLDE is President and CEO of the Partnership for New York City, a nonprofit organization of the city's business leaders, established by David Rockefeller in 1979. The Partnership is dedicated to maintaining New York City as a center of world commerce, finance, and innovation. Its public policy focus is on issues in the areas of education, infrastructure, and the economy.

The Partnership's economic development arm is the New York City Investment Fund. Wylde served as founding President and CEO of this $110 million civic fund, which was established in 1996 under the leadership of Henry R. Kravis. She continues to serve on the Fund's Board of Directors.

Wylde was also founding President and CEO of the Housing Partnership Development Corporation, serving from 1982-1996. In that capacity, she was instrumental in the creation of a number of pioneering initiatives in affordable housing at the local, state, and national levels. Under her leadership, more than $2 billion in private funds were invested in public-private partnerships that produced affordable housing and commercial developments in economically distressed communities across the city.

Wylde resides in Brooklyn and has a second home in Puerto Rico. She is a native of Madison, Wisconsin, and a graduate of St. Olaf College, 1968.

ROBERT D. YARO is the President of the Regional Plan Association, where he has been on the staff since 1990. Headquartered in Manhattan and founded in 1922, RPA is America's oldest and most respected independent metropolitan research and advocacy group.

Yaro also is Practice Professor in City and Regional Planning at the University of Pennsylvania. Formerly he served on the faculties of Harvard University and the University of Massachusetts at Amherst. Currently, he chairs the Civic Alliance to Rebuild Downtown New York, a broad-based coalition of civic groups formed to guide redevelopment in Lower Manhattan in the aftermath of the September 11, 2001 attacks on the World Trade Center. He is also a director of the Alliance for Regional Stewardship.

Yaro holds a Masters degree in City and Regional Planning from Harvard University and a B.A. in Urban Studies from Wesleyan University.

CONFERENCE SCHEDULE

Tuesday, September 5

4–8 p.m.   **Conference Registration**
Pace University, Mulitpurpose
Room

9–10 p.m.   **NY1 Town Hall Meeting,
with John Schiumo,
WTC Reconstruction**
The Call anchor John Schiumo
hosts a discussion on attempts
to re-build the WTC site. It will
look at the frustrations of the
last five years and what prom
ise future plans may hold.

Wednesday, September 6

8–9:30 a.m.   **Crain's New York Business
Breakfast Forum, Pace
University**

Featuring: Daniel L. Doctoroff, Deputy
Mayor for Economic
Development and Rebuilding,
New York City
Location: Multipurpose Room
and Schimmel Lobby
Networking Breakfast 8-8:30
a.m. Program 8:30-9:30 a.m.

10 a.m.–Noon   **Additional Conference
Registration**

Exhibition on Display
"The First 24 Hours" a New
York State Museum exhibition,
featuring a historic 40-foot

timeline and artifacts from the World Trade Center recovery will be on display in the Michael Schimmel Theater Lobby.

Noon–1:30 p.m.

Opening Keynote: David Gergen, Editor-at-Large, *US News & World Report*

PREPAREDNESS AND RESPONSE

1:30–2:45 p.m.

Panel 1: How has 9-11 Changed the Preparedness of First Responders: Are We Ready for the Next Attack?

Featuring: Jim Dwyer, reporter, NY *Times*, and author, *102 Minutes, The Untold Story of the Fight to Survive Inside the Twin Towers*; Michael N. Emmerman, Director, The Special Operations Support Group; Edward Galea, Director, Fire Safety Engineering, The University of Greenwich, U.K.; Joseph F. Ryan, Professor of Criminal Justice, Pace University, Panel Moderator

3–4 p.m.

Panel 2: Economic Impact: Global Business Community
Featuring: Joseph R. Baczko, Dean, Lubin

School of Business, Pace
University, Panel Moderator;
Eric Deutsch, President,
Alliance for Downtown New
York; Michael Dolfman,
Regional Commissioner, United
States Bureau of Labor
Statistics; Joseph T. Petro,
Executive Vice President and
Managing Director, Citigroup
Security and Investigative
Services, Panel Opener; Steven
Spinola, President, Real Estate
Board of New York

ENVIRONMENT/HEALTH

4:15–5:15 p.m.

**Panel 3:
Environmental
Consequences: Public
Health**

Featuring: John Cronin,
Director, Pace University
Academy for the Environment,
Panel Moderator; Dr. Bruce
Logan, President and CEO,
New York Downtown Hospital;
Jerrold Nadler, New York
Congressman; David M.
Newman, Industrial
Hygienist, New York
Committee for Occupational
Safety and Health; Lorna
Thorpe, Deputy Commissioner,
NYC Department of Mental
Health and Hygiene, Division

of Epidemiology, Panel Opener.

6 p.m.

***Project Rebirth* Reception & Preview of the Feature Film**

Featuring: Jim Whitaker, President, Imagine Entertainment
Screening 7-8 p.m.
www.projectrebirth.org

9–10 p.m.

NY1 Town Hall Meeting with Kristen Shaugnessy, WTC Health Concerns
NY1 News anchor Kristen Shaughnessy hosts a discussion on the continuing concerns over the health of residents and first responders as a result of the explosions and collapse of the twin towers.

Thursday, September 7

SURVIVORS

9–10 a.m.

Conference Registration
Michael Schimmel Theater Lobby

10:00–11:15 a.m.

Panel 4: If You Lived Here
Featuring: James Cavanaugh, President and CEO, Hugh L. Carey Battery Park City Authority, Panel Moderator and Opener; Thomas Healy, President, Lower Manhattan

Cultural Council; Charles Lai, Executive Director, Museum of Chinese in the Americas; Julie Menin, Chairperson, Community Board 1; Mark Schaming, Director of Exhibitions and Programs, Curator, New York State Museum; "The First 24 Hours" Exhibition

11:30–12:30 p.m.   **Panel 5: The Victims' Families and Their Influence on Public Policy**

Featuring: Sally Regenhard, Founder and Chairperson of the Skyscraper Safety Campaign; Tom Roger, Co-founder, Families of Sept. 11, Inc., and Director, WTC Memorial Foundation; Richard Shadick, Director, Counseling Center, Pace University, Panel Moderator; Eugene Steuerle, Co-Founder, Our Voices Together, Panel Opener

12:30–1:30 p.m.   **BREAK**

1:30–3 p.m.   **Keynote Address and Q&A: Lee Hamilton, Vice Chair, 9-11 Commission, and Member of the President's Homeland Security Advisory Council**

3:15–4:45 p.m.

Panel 6: Rebuilding, Repair and Hope

Featuring: John Cahill, Secretary to Governor George Pataki, Panel Opener; Stefan Pryor, Former President, Lower Manhattan Development Corporation; William C. Rudin, Chairman, Association for a Better New York; Kathryn Wylde, President, The Partnership for New York City, Panel Moderator; Robert Yaro, President, Regional Plan Association

5–6 p.m.

Panel 7: Higher Education
Featuring: David A. Caputo, President, Pace University; Robert Hackman, Citizen of Ghana, Graduate Student, Pace University; John Merrow, Veteran Producer, News Hour with Jim Lehrer, PBS Radio, Panel Moderator; Sadie Bragg, Provost, Borough of Manhattan Community College; David Warren, President, National Association of Independent Colleges and Universities, Panel Opener

9–10 p.m.

NY1 Town Hall Meeting with Budd Mishkin, Post

9/11 Security vs. Personal Liberties
NY1 News reporter Budd Mishkin hosts a discussion on a changed New York. What have we given up in terms of personal freedoms to guarantee our safety and security?

Friday, September 8
7:30–8:30 a.m.

Conference Registration
Michael Schimmel Theater Lobby

8:30–10 a.m.

Opening Keynote: Doris Kearns Goodwin, Author and Presidential Historian

POLITICS

10:15–11:30 a.m.

Panel 8: America's Place in the World
Featuring: Beverly Kahn, Vice President of International Opportunities, Pace University, Panel Moderator; Steven C. Clemons, Director, American Strategy Program, New America Foundation; Alice M. Greenwald, Executive Vice-President for Programs and Director of Memorial Museum, World Trade Center Memorial Foundation; Kenneth Jackson, Professor, Columbia University, Panel Opener; Nikki Stern, Former Executive Director of Families of September 11 and

Author, *Whose Moral Authority Is It Anyway?*

Noon–1:30 p.m.   **Conference Closing Remarks: William Kristol, Editor, *The Weekly Standard*, and Co-Author of *The War Over Iraq: America's Mission and Saddam's Tyranny***

9–10 p.m.   **NY1 Town Hall Meeting with Roma Torre, 9/11 and the Popular Arts**
NY1 News anchor Roma Torre hosts a discussion on the popular arts and 9/11. How has film, theatre, literature, and the arts in general, portrayed and interpreted the events of 9/11?

www.ingramcontent.com/pod-product-compliance
Lightning Source LLC
Chambersburg PA
CBHW070354270326
41926CB00014B/2542